Benchmarking of Control Strategies for Wastewater Treatment Plants

Scientific and Technical Report Series No. 23

Benchmarking of Control Strategies for Wastewater Treatment Plants

Edited by:

Krist V. Gernaey, Ulf Jeppsson, Peter A. Vanrolleghem and John B. Copp

IWA PUBLISHING

Published by **IWA Publishing**
Alliance House
12 Caxton Street
London SW1H 0QS, UK
Telephone: +44 (0)20 7654 5500
Fax: +44 (0)20 7654 5555
Email: publications@iwap.co.uk
Web: www.iwapublishing.com

First published 2014

British Library Cataloguing in Publication Data
A CIP catalogue record for this book is available from the British Library.

ISBN 9781843391463 (Paperback)
ISBN 9781780401171 (eBook)

Contents

Nomenclature

The following lists of symbols, abbreviations and acronyms are commonly used in this STR. Symbols specific to the various model variables and parameters are also listed in the appendices with their default values.

SYMBOLS

COD_{part}	chemical oxygen demand, particulates ($g\ COD \cdot m^{-3}$)
COD_{sol}	chemical oxygen demand, soluble material ($g\ COD \cdot m^{-3}$)
COD_{tot}	total chemical oxygen demand ($COD_{part} + COD_{sol}$) ($g\ COD \cdot m^{-3}$)
e	error of the controlled variable (units depend on variable)
K_La	oxygen mass transfer coefficient (d^{-1})
M	mass (kg)
N_{tot}	total nitrogen ($g\ N \cdot m^{-3}$)
N_{xx}	nitrogen content of variable *xx* ($kmol\ N \cdot (kg\ COD)^{-1}$)
$\mathbf{P}$	Markov chain probability matrix (–)
P_{atm}	atmospheric pressure (bar)
P_{gas}	gas pressure (bar)
p_{ij}	transition probability from state *i* to state *j* in a Markov chain (–)
$p_{gas,xx}$	partial pressure of gas type *xx* (bar)
Q_{av}	average flow rate ($m^3 \cdot d^{-1}$)
Q_{do}	overflow from dewatering, reject water ($m^3 \cdot d^{-1}$)
Q_e	effluent flow rate ($m^3 \cdot d^{-1}$)
Q_{EC}	external carbon flow rate ($m^3 \cdot d^{-1}$)
Q_f	feed flow rate ($m^3 \cdot d^{-1}$)
Q_{gas}	gas flow rate ($Nm^3 \cdot d^{-1}$)
Q_i and Q_{in}	influent flow rate ($m^3 \cdot d^{-1}$)
Q_{int}	internal recycle flow rate ($m^3 \cdot d^{-1}$)

Q_{po}	overflow from primary clarifier ($m^3 \cdot d^{-1}$)
Q_{pu}	underflow from primary clarifier ($m^3 \cdot d^{-1}$)
Q_r	return sludge flow rate ($m^3 \cdot d^{-1}$)
Q_{so}	overflow from secondary clarifier ($m^3 \cdot d^{-1}$)
Q_{su}	underflow from secondary clarifier ($m^3 \cdot d^{-1}$)
Q_{to}	overflow from thickener ($m^3 \cdot d^{-1}$)
Q_w	flow rate of waste sludge ($m^3 \cdot d^{-1}$)
R	universal gas constant (0.083145 bar $\cdot$ (kmol $\cdot$ $m^{-3})^{-1} \cdot K^{-1}$)
$R_{td/tr}$	ratio of delay time to response time of a sensor (–)
S_{e95}	95 percentile of a soluble substance (units depend on variable)
S_0^{sat}	saturated oxygen concentration (g $O_2 \cdot m^{-3}$)
S_{tox}	toxic substance in soluble form (g $\cdot$ m^{-3})
t	time (d)
T	temperature (°C)
t_d	delay time (sensor models) (min)
T_d	derivative time in a PID controller (d)
t_i	measurement interval (sensor models) (min)
T_i	integral time in a PID controller (d)
t_{obs}	integration interval for integral (d)
t_r	response time (min)
TSS_e	effluent total suspended solids concentration (g SS $\cdot$ m^{-3})
u	value of the manipulated variable or input signal (units depend on variable)
V	volume (m^3)
V_{aer}	aerobic volume of plant (m^3)
X_{e95}	95 percentile of a particulate substance (units depend on variable)
X_{tox}	toxic substance in particulate form (g $\cdot$ m^{-3})
y	output signal (units depend on variable)
β	weighting factor for effluent quality (units depend on variable)
δ	standard deviation of measurement noise (units depend on variable)
η_{CODp}	particulate COD removal efficiency of primary clarifier (–)
Δt	time interval between controller executions (d)
τ	time constant for sensors and actuators (d)

ASM1 symbols (including secondary clarifier)

b_A	autotrophic decay rate (d^{-1})
b_H	heterotrophic decay rate (d^{-1})
f_{ns}	non-settleable fraction (–)
f_p	fraction of biomass leading to particulate inert products (–)
i_{XB}	fraction of nitrogen in biomass (g N $\cdot$ (g COD)$^{-1}$)
i_{XP}	fraction of nitrogen in organic particulate inerts (g N $\cdot$ (g COD)$^{-1}$)
K_{NH}	ammonia half-saturation coefficient for autotrophic growth (g N $\cdot$ m^{-3})
K_{NO}	nitrate half-saturation coefficient for anoxic heterotrophic growth (g N $\cdot$ m^{-3})
K_{OA}	oxygen half-saturation coefficient for autotrophic growth (g $O_2 \cdot m^{-3}$)
K_{OH}	oxygen half-saturation coefficient for heterotrophic growth (g $O_2 \cdot m^{-3}$)
K_S	substrate half-saturation coefficient for heterotrophic growth (g COD $\cdot$ m^{-3})

K_X	particulate substrate half-saturation coefficient for hydrolysis (g COD · (g COD)$^{-1}$)
k_a	ammonification rate (m^3 · (g COD)$^{-1}$ · d^{-1})
k_H	maximum specific hydrolysis rate (g COD · (g COD)$^{-1}$ · d^{-1})
r_h	hindered zone settling parameter (m^3 · g^{-1})
r_p	flocculant zone settling parameter (m^3 · g^{-1})
S_{ALK}	alkalinity (mol HCO_3^- · m^{-3})
S_I	soluble inert organic matter (g COD · m^{-3})
S_{ND}	soluble biodegradable organic nitrogen (g N · m^{-3})
S_{NH}	ammonium plus ammonia nitrogen (g N · m^{-3})
S_{NO}	nitrate and nitrite nitrogen (g N · m^{-3})
S_O	dissolved oxygen (g O_2 · m^{-3})
S_S	readily biodegradable substrate (g COD · m^{-3})
$X_{B,A}$	active autotrophic biomass (g COD · m^{-3})
$X_{B,H}$	active heterotrophic biomass (g COD · m^{-3})
X_I	particulate inert organic matter (g COD · m^{-3})
X_{ND}	particulate biodegradable organic nitrogen (g N · m^{-3})
X_P	particulate products arising from biomass decay (g COD · m^{-3})
X_S	slowly biodegradable substrate (g COD · m^{-3})
X_{min}	minimum attainable solids concentration in secondary clarifier (g SS · m^{-3})
Y_A	autotrophic yield (g COD · (g N)$^{-1}$)
Y_H	heterotrophic yield (g COD · (g COD)$^{-1}$)
η_g	anoxic growth rate correction factor (–)
η_h	anoxic hydrolysis rate correction factor (–)
μ_{mA}	maximum autotrophic growth rate (d^{-1})
μ_{mH}	maximum heterotrophic growth rate (d^{-1})
v_o	maximum Vesilind settling velocity (m · d^{-1})
v'_o	maximum settling velocity (m · d^{-1})
v_s	settling velocity (m · d^{-1})

ADM1 Symbols (including most model parameters)

C_{aa}	carbon content of S_{aa} (kmol C · (kg COD)$^{-1}$)
C_{ac}	carbon content of S_{ac} (kmol C · (kg COD)$^{-1}$)
C_{bac}	carbon content of biomass (kmol C · (kg COD)$^{-1}$)
C_{bu}	carbon content of S_{bu} (kmol C · (kg COD)$^{-1}$)
C_{ch}	carbon content of X_{ch} (kmol C · (kg COD)$^{-1}$)
C_{ch4}	carbon content of S_{ch4} (kmol C · (kg COD)$^{-1}$)
C_{fa}	carbon content of S_{fa} (kmol C · (kg COD)$^{-1}$)
C_{li}	carbon content of X_{li} (kmol C · (kg COD)$^{-1}$)
C_{pr}	carbon content of X_{pr} (kmol C · (kg COD)$^{-1}$)
C_{pro}	carbon content of S_{pro} (kmol C · (kg COD)$^{-1}$)
C_{SI}	carbon content of S_I (kmol C · (kg COD)$^{-1}$)
C_{su}	carbon content of S_{su} (kmol C · (kg COD)$^{-1}$)
C_{va}	carbon content of S_{va} (kmol C · (kg COD)$^{-1}$)
C_{xc}	carbon content of X_c (kmol C · (kg COD)$^{-1}$)
C_{XI}	carbon content of X_I (kmol C · (kg COD)$^{-1}$)
$f_{ac,aa}$	yield (catabolism only) of S_{ac} on S_{aa} (–)

$f_{ac,su}$	yield (catabolism only) of S_{ac} on S_{su} (–)
$f_{bu,aa}$	yield (catabolism only) of S_{bu} on S_{aa} (–)
$f_{bu,su}$	yield (catabolism only) of S_{bu} on S_{su} (–)
$f_{ch,xc}$	fraction of composites to X_{ch} by disintegration (–)
$f_{fa,li}$	yield (catabolism only) of S_{fa} on X_{li} (–)
$f_{h2,aa}$	yield (catabolism only) of S_{h2} on S_{aa} (–)
$f_{h2,su}$	yield (catabolism only) of S_{h2} on S_{su} (–)
$f_{li,xc}$	fraction of composites to X_{li} by disintegration (–)
$f_{pr,xc}$	fraction of composites to X_{pr} by disintegration (–)
$f_{pro,aa}$	yield (catabolism only) of S_{pro} on S_{aa} (–)
$f_{pro,su}$	yield (catabolism only) of S_{pro} on S_{su} (–)
$f_{SI,xc}$	fraction of composites to S_I by disintegration (–)
$f_{XI,xc}$	fraction of composites to X_I by disintegration (–)
$f_{va,aa}$	yield (catabolism only) of S_{va} on S_{aa} (–)
$K_{a,acid}$	acid-base equilibrium constant for *acid* ($kmol \cdot m^{-3}$)
$k_{A,Bsub}$	acid-base kinetic parameter for substance *sub* ($m^3 \cdot kmol^{-1} \cdot d^{-1}$)
$k_{dec,bac}$	decay rate for bacteria of type *bac* (d^{-1})
k_{dis}	disintegration rate (d^{-1})
$K_{H,gas}$	Henry's law coefficient for type of *gas* ($kmol \cdot m^{-3} \cdot bar^{-1}$)
$k_{hyd,ch}$	hydrolysis rate of X_{ch} (d^{-1})
$k_{hyd,li}$	hydrolysis rate of X_{li} (d^{-1})
$k_{hyd,pr}$	hydrolysis rate of X_{pr} (d^{-1})
$K_{I,inhib,proc}$	50% inhibitory concentration of inhibitor *inhib* on process *proc* ($kg\ COD \cdot m^{-3}$)
$k_{m,proc}$	Monod maximum specific uptake rate for process *proc* (d^{-1})
$K_{S,proc}$	half saturation value for process *proc* ($kg\ COD \cdot m^{-3}$)
N_{xc}	nitrogen content of X_c ($kmol\ N \cdot (kg\ COD)^{-1}$)
N_{aa}	nitrogen content of S_{aa} ($kmol\ N \cdot (kg\ COD)^{-1}$)
N_{bac}	nitrogen content of biomass ($kmol\ N \cdot (kg\ COD)^{-1}$)
N_I	nitrogen content of inerts ($kmol\ N \cdot (kg\ COD)^{-1}$)
S_{aa}	amino acids ($kg\ COD \cdot m^{-3}$)
S_{ac}	total acetate, sum of acid-base pairs ($kg\ COD \cdot m^{-3}$)
S_{an}	anions ($kmol \cdot m^{-3}$)
S_{bu}	total butyrate, sum of acid-base pairs ($kg\ COD \cdot m^{-3}$)
S_{cat}	cations ($kmol \cdot m^{-3}$)
S_{ch4}	methane ($kg\ COD \cdot m^{-3}$)
S_{fa}	long chain fatty acids ($kg\ COD \cdot m^{-3}$)
S_{h2}	hydrogen ($kg\ COD \cdot m^{-3}$)
S_I	soluble inerts ($kg\ COD \cdot m^{-3}$)
S_{IC}	inorganic carbon, sum of acid-base pairs ($kmol\ C \cdot m^{-3}$)
S_{IN}	inorganic nitrogen, sum of acid-base pairs ($kmol\ N \cdot m^{-3}$)
S_{pro}	total propionate, sum of acid-base pairs ($kg\ COD \cdot m^{-3}$)
S_{su}	monosaccharides ($kg\ COD \cdot m^{-3}$)
S_{va}	total valerate, sum of acid-base pairs ($kg\ COD \cdot m^{-3}$)
X_{aa}	biomass, amino acid degraders ($kg\ COD \cdot m^{-3}$)
X_{ac}	biomass, acetate degraders ($kg\ COD \cdot m^{-3}$)
X_c	composites ($kg\ COD \cdot m^{-3}$)

X_{c4}	biomass, valerate and butyrate degraders (kg COD · m^{-3})
X_{ch}	carbohydrates (kg COD · m^{-3})
X_{fa}	biomass, long chain fatty acids degraders (kg COD · m^{-3})
X_{h2}	biomass, hydrogen degraders (kg COD · m^{-3})
X_I	particulate inerts (kg COD · m^{-3})
X_{li}	lipids (kg COD · m^{-3})
X_{pr}	proteins (kg COD · m^{-3})
X_{pro}	biomass, propionate degraders (kg COD · m^{-3})
X_{su}	biomass, sugar degraders (kg COD · m^{-3})
Y_{aa}	yield of biomass, amino acid degraders (kg COD · (kg COD)$^{-1}$)
Y_{ac}	yield of biomass, acetate degraders (kg COD · (kg COD)$^{-1}$)
Y_{c4}	yield of biomass, valerate and butyrate degraders (kg COD · (kg COD)$^{-1}$)
Y_{fa}	yield of biomass, long chain fatty acid degraders (kg COD · (kg COD)$^{-1}$)
Y_{h2}	yield of biomass, hydrogen degraders (kg COD · (kg COD)$^{-1}$)
Y_{pro}	yield of biomass, protein degraders (kg COD · (kg COD)$^{-1}$)
Y_{su}	yield of biomass, sugar degraders (kg COD · (kg COD)$^{-1}$)

ABBREVIATIONS AND ACRONYMS

AD	anaerobic digestion
ADM1	Anaerobic Digestion Model No. 1
AE	aeration energy
AS	activated sludge
ASM1	Activated Sludge Model No. 1
ATV	German Association for Water Economy, Wastewater and Waste
BOD	biochemical oxygen demand
BOD_5	biochemical oxygen demand measured over 5 days
BSM(s)	benchmark simulation model(s)
BSM1	Benchmark Simulation Model No. 1
BSM1_LT	Benchmark Simulation Model No. 1 Long-Term
BSM2	Benchmark Simulation Model No. 2
C	Celsius
CL	closed loop
COD	chemical oxygen demand
COST	European Co-Operation in the field of Scientific and Technical research
CV	controlled variable
DAE	differential algebraic equation
DIPDSG	dynamic influent pollution disturbance scenario generator
DO	dissolved oxygen
DW	dewatering unit
EC	external carbon
EPA	Environmental Protection Agency
EQ	effluent quality
EQI	effluent quality index
EU	European Union
FAC	false acceptance

FAL	false alarm
HE	heating energy
HRT	hydraulic retention time
IAE	integral of absolute error
ICA	instrumentation, control and automation
IQ	influent quality
IQI	influent quality index
ISE	integral of squared error
IWA	International Water Association
K	Kelvin
ME	mixing energy
MIA	modelling and integrated assessment
MLE	modified Ludzack-Ettinger (plant configuration)
MLSS	mixed liquor suspended solids
MLVSS	mixed liquor volatile suspended solids
MP	methane production
MV	manipulated variable
N	nitrogen
ODE	ordinary differential equation
OL	open loop
OCI	operational cost index
OUR	oxygen uptake rate
OTR	oxygen transfer rate
P	phosphorus
PE	pumping energy
peq	person equivalent
pH	$-\log_{10}[H^+]$
PI	proportional – integral (controller type)
PID	proportional – integral – derivative (controller type)
PRIM	primary clarifier
PU	pollution unit
RAS	return activated sludge
SC	secondary clarifier
SOTE	standard oxygen transfer efficiency
SP	sludge production
SRT	solids retention time
SS	suspended solids
ST	storage tank
STR	Scientific and Technical Report
TG	Task Group
THK	thickener
TKN	total Kjeldahl nitrogen
TN	total nitrogen

TR	technical report (generated as part of this work)
TSS	total suspended solids
VFA	volatile fatty acids
VSS	volatile suspended solids
WAS	waste activated sludge
WWC	world water congress
WWTP(s)	wastewater treatment plant(s)

List of technical reports

This Scientific and Technical Report (STR) is a comprehensive summary of nearly 20 years of effort. Over the years, fifteen (15) detailed technical reports on specific aspects of the work have been generated and these are listed below. These fifteen technical reports document the most detailed information provided by the Task Group and should be considered appendices or associated documents to this STR. Furthermore, it is the intent of the Task Group to maintain these reports as living documents, such that as new related information becomes available in the future, these reports will be updated. All of the reports are electronically available to those that purchase the STR.

1. Pons M.-N., Alex J., Benedetti L., Copp J. B., Gernaey K. V., Jeppsson U., Nopens I., Rosen C., Steyer J.-P. and Vanrolleghem P. A. (2014). Benchmark Simulation Model No. 1 (BSM1). Technical Report No. 1. IWA Task Group on Benchmarking of Control Strategies for Wastewater Treatment Plants.
2. Alex J., Benedetti L., Copp J., Corominas L., Gernaey K. V., Jeppsson U., Nopens I., Pons M.-N., Rosen C., Steyer J.-P. and Vanrolleghem P. A. (2014). Long-Term Benchmark Simulation Model No. 1 (BSM1_LT). Technical Report No. 2. IWA Task Group on Benchmarking of Control Strategies for Wastewater Treatment Plants.
3. Alex J., Benedetti L., Copp J., Gernaey K. V., Jeppsson U., Nopens I., Pons M.-N., Rosen C., Steyer J.-P. and Vanrolleghem P. A. (2014). Benchmark Simulation Model No. 2 (BSM2). Technical Report No. 3. IWA Task Group on Benchmarking of Control Strategies for Wastewater Treatment Plants.
4. Gernaey K. V., Alex J. and Copp J. (2014). Primary Clarifier Model for BSM2 Application. Technical Report No. 4. IWA Task Group on Benchmarking of Control Strategies for Wastewater Treatment Plants.
5. Rosen C. and Jeppsson U. (2014). Aspects on ADM1 Implementation within the BSM2 Framework. Technical Report No. 5. IWA Task Group on Benchmarking of Control Strategies for Wastewater Treatment Plants.
6. Alex J., Corominas L., Rieger L. and Winkler S. (2014). Models of Sensors and Actuators in the BSM Framework. Technical Report No. 6. IWA Task Group on Benchmarking of Control Strategies for Wastewater Treatment Plants.

7. Corominas L., Jeppsson U., Vanrolleghem P. A. (2014). Ring-Testing of Benchmark Sensor Models. Technical Report No. 7. IWA Task Group on Benchmarking of Control Strategies for Wastewater Treatment Plants.
8. Gernaey K. V., Flores-Alsina X., Benedetti L., Rosen C. and Jeppsson U. (2014). A Model for Dynamic Influent Data Generation. Technical Report No. 8. IWA Task Group on Benchmarking of Control Strategies for Wastewater Treatment Plants.
9. Vanrolleghem P. A., Batstone D. and Pons M.-N. (2014). Temperature Effects in BSM1_LT and BSM2. Technical Report No. 9. IWA Task Group on Benchmarking of Control Strategies for Wastewater Treatment Plants.
10. Nopens I., Alex J., Batstone D., Copp J., Dudley J., Pons M.-N., Vanrolleghem P. A, Volcke E. I. P., and Jeppsson U. (2014). ASM/ADM/ASM Interfaces for BSM2. Technical Report No. 10. IWA Task Group on Benchmarking of Control Strategies for Wastewater Treatment Plants.
11. Comas J. and Rodríguez-Roda I. (2014). Risk Model for Microbiology-Related Settling Problems. Technical Report No. 11. IWA Task Group on Benchmarking of Control Strategies for Wastewater Treatment Plants.
12. Comas J. and Rodríguez-Roda I. (2014). Implementation of Risk Model for Microbiology-Related Settling Problems. Technical Report No. 12. IWA Task Group on Benchmarking of Control Strategies for Wastewater Treatment Plants.
13. Vrecko D., Gernaey K. V., Jeppsson U., Volcke E. I. P. and Vanrolleghem P. A. (2014). Evaluation Criteria Description and Examples for BSM2. Technical Report No. 13. IWA Task Group on Benchmarking of Control Strategies for Wastewater Treatment Plants.
14. Nopens I., De Keyser W., Alex J., Amerlinck Y., Benedetti L., Copp J., Corominas L., Dudley J., Pons M.-N., Rosen C., Vanrolleghem P.A. and Jeppsson U. (2014). Ring-Testing of BSM1 and BSM2. Technical Report No. 14. IWA Task Group on Benchmarking of Control Strategies for Wastewater Treatment Plants.
15. Jeppsson U. and Vanrolleghem P. A. (2014). Publications Associated with the Benchmark Simulation Models. Technical Report No. 15. IWA Task Group on Benchmarking of Control Strategies for Wastewater Treatment Plants.

Preface

The simulation-based benchmarking of wastewater treatment plant (WWTP) control strategies was first suggested 21 years ago at the Instrumentation, Control and Automation conference in Hamilton, Canada and this STR represents another milestone in that development. Thanks to a unique, comprehensive, long-term international collaboration on a mostly voluntary basis, this is the third major milestone in the overall development of this tool that permits the objective simulation-based evaluation of the performance of process control and monitoring strategies. The well-known COST-682 report (Copp, 2002) described the first benchmark simulation protocol and the IWA Scientific and Technical Report No. 11 (Copp *et al.* 2002) summarised the work on benchmarking respirometry-based control strategies in WWTPs similar to the COST-682-plant, but the work reported here has moved the benchmarking idea much further. This work describes both spatial (whole plant control performance evaluation) and temporal extensions (performance evaluation over a one year period) for not only the evaluation of controller performance, but also process monitoring performance including the evaluation of algorithms for fault detection in sensors and actuators. And, this development only seems to be the beginning as numerous new ideas have surfaced in post-project audits of the benchmarking development, as presented by Jeppsson *et al.* (Watermatex2011, San Sebastian, Spain), Gernaey *et al.* (ICA2013, Narbonne, France) and Vanrolleghem *et al.* (WWTmod2014, Spa, Belgium).

With over 400 publications (a number that keeps climbing) using the benchmark or parts of it, there is no doubt that the work has been embraced by the scientific community. But, while the original objective of producing a tool to objectively compare control strategies may have been less successful than anticipated, the range of indirect benefits that have emerged from the collaborative effort has had a huge impact in the WWTP modelling profession. And, thanks to the support from IWA and the COST programme, this material has, for the most part, been made available to the research community for free, which has contributed to its success. Scrutinised model structures, verified model implementations in different simulation platforms, structured thinking around process and control performance evaluation criteria, influent generators and course material are what the profession has appreciated. This comprehensive effort has made the benchmark simulation models an ideal source of material for professors, PhD students and other researchers all over the world. Without the need to implement and test their platform, these beneficiaries have saved countless hours that presumably have been diverted to answering important research questions rather than wasting time debugging the various models. In this way, the Task Group (TG) is convinced that the benchmark has, among other things promoted innovation by providing researchers with additional time to be creative and to use the models, instead of spending time repeating the obvious.

Looking back, when the EU's COST-action support (probably the most cost-effective EU research support tool ever) of the benchmark work came to an end (Actions 624 and 682), it left a large group of enthusiastic benchmarkers without a means to meet and discuss further benchmarking developments. This came to a head in Marrakech (Morocco, IWA World Water Congress 2004) when the formation of a benchmarking Task Group was discussed with IWA officials. The response was positive and in August 2005 the TG was officially launched. The Task Group work was supposed to take three years and final outcomes were to be presented at a workshop at IWA's WWC 2008 (Vienna, Austria). However, it became clear very early on that the scope of work exceeded the Task Group's original estimate and as a result, it clearly took more time to complete than originally proposed (sincere apologies are in order to all that have been waiting for this STR to be published). The fact that the core group members took on different responsibilities over time did not help and many in that core group now spend their days coordinating research rather than doing hands-on research themselves.

It is time to pass on the responsibility for developing the benchmark models further to the next generation of young researchers and this Task Group is confident that those individuals will continue to promote cooperation and the free exchange of modelling ideas in an effort to form a consensus and move the science of modelling forward for the benefit of all.

It is also time for a great many thanks. The four editors of this STR should be considered spokespersons of a worldwide endeavour in which we estimate over a hundred people have contributed to reach the result you are reading. We are fortunate to have been able to make use of supporting groups, such as IWA Specialist Groups on Modelling and Integrated Assessment (MIA) and Instrumentation, Control and Automation (ICA). Workshops were organised during which stepwise progress was made and homework was delegated to the large group of MSc and PhD students and collaborators that were active within the wider benchmarking development team. The COST-682 and -624 actions allowed tens of people to converge on meeting places in Europe two to three times a year, and IWA Task Group meetings at larger conferences were instrumental in providing a development platform and fostering friendships that will forever outlast the development.

It is dangerous to provide an explicit list of contributors, but we are not risk-averse and hope that any person feeling he/she has been forgotten, will forgive us for making a mistake, and understand it is important to try to personally thank as many as possible. However, let us first start by acknowledging the leadership of the other members of the Task Group's core group:

Jens Alex *Marie-Noëlle Pons* *Christian Rosén* *Jean-Philippe Steyer*

Secondly, we want to acknowledge the tremendous efforts of all other co-authors of this STR and the associated technical reports:

Youri Amerlinck	*Lluis Corominas*	*Ingmar Nopens*	*Eveline Volcke*
Damien Batstone	*Webbey De Keyser*	*Leiv Rieger*	*Darko Vrecko*
Lorenzo Benedetti	*Jeremy Dudley*	*Ignasi Rodríguez-Roda*	*Stefan Winkler*
Joaquim Comas	*Xavier Flores-Alsina*		

And also many thanks to all other contributors who, in one way or another, have influenced and/or worked on the BSM development over the years:

Abdallah Abusam
Daniel Aguado
Robert Andersson
Magnus Arnell
Eduardo Ayesa
Ouadiaa Barrou
Farid Benazzi
Benoit Beraud
Davide Bixio
Bengt Carlsson
Josep Carrasco Martínez
Magali Casellas
Filip Claeys
Petra Claeys
Jean-Pierre Corriou
Christophe Dagot
Jordi Dalmau
Mónica de Gracia
Dirk De Pauw
Lieven De Temmerman
Joel Denamur
Martijn Devisscher
Stefan Diehl
René Dupont
George Ekama
Mats Ekman
Sebastian Farås
Samo Gerksic
Jairo Gómez
Paloma Grau
Javier Guerrero
Lisha Guo
Chris Hellinga
Mats Holmberg
Estibaliz Huete
Nadja Hvala
Malin Jonasson
Ulrich Jumar
Karel Keesman
Günter Langergraber
Cyril Lemoine
Yann Lemoullec
Mathieu Lesteur
Erik Lindblom
M.d.C. Lourenço da Silva
Thomas Maere
Stefano Marsili-Libelli
Cristina Martin
Romain Martin
Michael Ogurek
Gustaf Olsson
Phuong Thu Pham
Manel Poch
José Porro
Olivier Potier
Yang Qing
Isabelle Queinnec
Botond Raduly
Ivan Ramirez
Ferran Ribas
Josep Ribes
Vicky Ruano
Magda Ruiz-Ordónez
Joakim Rydh
Ramesh Saagi
Pär Samuelsson
Manfred Schütze
Gürkan Sin
Laura Snip
Kimberly Solon
Henri Spanjers
Mathieu Spérandio
Aljaz Stare
Imre Takács
Henk Vanhooren
Kris Villez
Bernhard Wett
Ma Yong
ChangKyoo Yoo
Usama Zaher
Ann Åkesson
Linda Åmand

Finally, we would be remiss if we did not acknowledge the tremendous effort of Professor Zhiguo Yuan of the University of Queensland, Australia, who not only contributed to the BSM development, but also graciously agreed to review and comment on the draft chapters. His efforts identified several areas of concern with respect to reader and concept understanding and gave us the opportunity to incorporate that feedback into the final manuscript. We are confident that the STR is now a better product because of Zhiguo's input and we are extremely thankful for his contribution.

We trust you will all enjoy and benefit from this STR!

Peter A. Vanrolleghem
Ulf Jeppsson
Krist V. Gernaey
John B. Copp
Québec, Lund, Lyngby and Hamilton

Chapter 1

Introduction

P. A. Vanrolleghem, J. B. Copp, K. V. Gernaey and U. Jeppsson

This Scientific and Technical Report (STR) is the summary of the work of the IWA Task Group on Benchmarking of Control Strategies for Wastewater Treatment Plants. As will be explained in Chapter 2, this Task Group has a long history. However, before describing this history and the results of the Task Group in more detail, we would first like to use this introduction to provide more insight into a number of basic issues related to the family of Benchmark Simulation Models (BSMs), which are the main 'products' of this Task Group. In order to do this, we will seek answers to a few basic questions: What is meant by a 'Benchmark Simulation Model'? What is the purpose of the BSMs? Who should use the BSMs? How should a BSM be used? Who has been involved in the development of the BSMs? And last but not least, how should this STR be read?

1.1 WHAT IS MEANT BY A 'BENCHMARK SIMULATION MODEL'?

When checking a dictionary, a benchmark is defined as a measure of reference to be used in a test. In computer science a benchmark is a reference performance to which the relative performance of hardware or software can be compared. In process modelling and control, a benchmark is defined as a plant model and associated control strategy that can be used as a reference point for simulation-based comparison of control strategies (Downs & Vogel, 1993). Such a simulation benchmark is not associated with a particular simulation platform. Direct coding (e.g., C/C++, FORTRAN) as well as commercial simulation software packages can be used. In this case, the purpose of the simulation protocol was to create a tool that could guarantee that different users obtain exactly the same results when running the simulation model. The main 'products' of this Task Group are WWTP simulation models (Chapter 4), a simulation protocol for these simulation models (Chapter 7) and a set of benchmarking evaluation criteria for objective control evaluation (Chapter 6). All these items together form the benchmark simulation model platform. However, it should be emphasised as well that a major result of the Task Group work is a set of verified unit process models and tools that are applicable to WWTP simulation studies in general. The Activated Sludge Model No. 1 (ASM1), Takács secondary clarifier model and Anaerobic Digestion Model No. 1 (ADM1) among others were all verified before including those unit process models in the BSM platform. More details on these models will be provided in Chapter 4. Other tools presented in Chapter 4, such as a dynamic influent pollution disturbance scenario generator (DIPDSG) model, AS-AD interface models, sensor and actuator (fault) models and a risk assessment module for microbiology related settling problems have also resulted from the work of the Task Group.

1.2 WHAT IS THE PURPOSE OF THE BENCHMARK SIMULATION MODELS?

The activated sludge process aims to achieve, at minimum cost, sufficiently low concentrations of biodegradable matter and nutrients in the effluent together with minimal sludge production. To do this, the process has to be controlled. Many control strategies have been proposed in the literature; however, the literature does not provide a clear basis for comparison of these strategies because of the many confounding influences that have an impact on the system. Many of these influences are easily recognised. For instance, physical characteristics of the process can have an impact on process performance and this makes the comparison of strategies applied to different reactor layouts difficult. As well, the influence of a control strategy on process performance is expected to vary with different disturbances, thus the disturbances used to test the control strategy become important. Also complicating the evaluation is the lack of standard evaluation criteria. Effluent requirements are for example often location specific, which makes it difficult to judge the particular influence of an applied control strategy from a reported performance increase. Controversies that result from the resulting subjective reports reinforce the need to devise an effective and unbiased evaluation method that can be used to judge the utility of different control strategies.

From a practical standpoint, it is not reasonable to experimentally test and verify the effectiveness of all reported control strategies and often the assessment of these control strategies is confounded by the multi-faceted nature of the process under study. Alternatively, given a standardised procedure, it is possible to efficiently evaluate numerous strategies through realistic/dynamic computer simulations. Simulations provide a cost-effective means for the evaluation of control strategies, but the unlimited number of simulation permutations makes the need for a standardised protocol very important if different strategies (and different simulation results) are to be compared. Each control strategy must be simulated under the same conditions to ensure unbiased comparisons. Validation of the computer simulations is difficult without supporting experimental or full-scale data, but the value of the work is enhanced through the use of accepted activated sludge models. Because appropriate simulation tools for the activated sludge process are available this approach has numerous advantages, but still there is a need for a standardised procedure. To this end, there has been an effort to develop a standardised simulation protocol – a 'simulation benchmark'.

The BSM platform was originally intended exclusively for the simulation-based comparison of WWTP control strategies. This was later extended to include the comparison of WWTP monitoring strategies as well (Chapters 4, 5, 6 and 7). Process monitoring is the activity by which collected data are analysed to find process deviations, failures and faults. Process monitoring also involves isolation of variables contributing to the deviations, facilitating further analysis of the problems. Again, many monitoring methods have been proposed in the literature but an unbiased evaluation of their performance in a WWTP was not available even though it was highly desired. Consequently, a standardised simulation protocol, verified models and evaluation criteria were established to address this deficiency.

1.3 WHO SHOULD USE THE BENCHMARK SIMULATION MODELS?

It is not that easy to describe a specific target group for the BSM platform. In fact, the models and tools described in this STR should appeal to a broad audience – that audience will be called 'benchmark users' in the rest of the STR – both from industry and academia with a general interest in WWTP modelling. This is best illustrated with a few examples. The benchmark simulation models can be used for objective simulation-based comparison of control strategies. As such, users from academia might find it interesting to develop a control strategy that yields the best overall performance. They may also use it to train their students on the basics of process control. A consultant, on the other hand, might use these models to demonstrate the use of control strategies to a potential customer. A benchmark user might also be interested in using only part of the benchmark simulation model (e.g., the evaluation criteria, or sensor and actuator models, or the influent generator). And of course, individual unit process models can be used in any simulation study as they all are now verified by the Task Group and are distributed for free when requested.

1.4 HOW SHOULD THE BENCHMARK SIMULATION MODELS BE USED?

The BSM platform is a standardised simulation protocol. If it is the intention of the benchmark user to work with the BSM platform and to compare simulation results with other benchmark users, then the standardised simulation protocol should be followed. A space limitation in this STR means that it was not possible to provide a description of all the minute details of the BSMs. Rather, this publication provides a detailed overview of the BSMs and the various tools. For more specific details, the reader is referred to the detailed technical reports and BSM computer code which accompany the STR.

The benchmark user is of course free to modify the models and tools provided by the Task Group. However, such modifications should be mentioned clearly and documented properly when publishing results in order to allow a fair comparison of results.

1.5 WHO HAS BEEN INVOLVED IN THE DEVELOPMENT OF THE BENCHMARK SIMULATION MODELS?

As will be explained in Chapter 2 it has taken several years to develop the BSM platform and the Task Group 'products' are truly the result of a group effort. Over the years, many people have contributed to the BSM platform development. These people were not all part of the Task Group and many did not specifically contribute to this publication. However, those individuals should be fully acknowledged for their work and input. The Task Group has tried to do so in the preface to this publication, knowing that we will probably never be able to name all the people that have been involved in the development over the years. In the rest of this STR, we will collectively refer to the BSM developers as the 'benchmarkers'.

1.6 HOW SHOULD THIS SCIENTIFIC AND TECHNICAL REPORT BE READ?

This STR is only a summary of the work of the Task Group and can be used to get a rapid overview of the results. The detailed technical reports were written by the benchmarkers while developing, implementing and validating the specific models and tools, and are as such the most detailed information provided by the Task Group. To supplement that reading, the reader can consult the model code provided by the Task Group in order to get a deeper understanding of the more technical issues. Finally, a benchmark user can find inspiration in the literature, as a significant number of scientific papers published during the years have been devoted to the use of the BSMs. These papers are listed in Technical Report No. 15. As a courtesy to the original contributors, benchmark users are asked to refer to those literature sources when publishing their own work, in order to properly acknowledge the work done by the different benchmarkers.

Chapter 2

Benchmark overview

M.-N. Pons, C. Rosen and U. Jeppsson

This chapter of the STR provides a concise historic overview of the key events and decisions that have resulted in the development of the different Task Group WWTP simulation benchmarks.

2.1 BENCHMARK SIMULATION MODEL NO. 1

In chemical engineering, the Tennessee-Eastman plant has been used extensively in research and teaching for process optimisation, process control and process monitoring. The plant is a process modelling and control benchmark describing a reactor/separator/recycle arrangement involving two main simultaneous gas-liquid exothermic reactions with associated operating objectives, process control objectives and process disturbances (Downs & Vogel, 1993).

Undoubtedly, the concepts behind the Tennessee-Eastman plant have been a major source of inspiration for the development of the WWTP simulation benchmarks described in this STR. In fact, the benchmarking idea was first mentioned by Professor Bengt Carlsson (Uppsala University, Sweden) at IWA's Instrumentation Control and Automation Conference in Hamilton (Canada) in 1993. This idea was inspired by the Control System Society in the USA related to flight control systems. The development of the first simulation benchmark (BSM1) was thus initiated two decades ago. The initial benchmarking concepts were largely defined in parallel with the work reported by Spanjers *et al.* (1998) and Copp *et al.* (2002), who developed and applied an activated sludge benchmark that focused on respirometry-based control. The concepts for this respirometry-based control benchmark were presented at a COST (European Cooperation in Science and Technology) 682 Working Group No. 2 meeting in Florence (Italy) in November 1996, and were very helpful as a starting point for the BSM1 development. A meeting in Wageningen (The Netherlands) in April 1997 was the true starting point for the BSM1 definition: the first model equations were written down and tasks to develop the main parts of BSM1 (performance evaluation, testing, influent definition, ...) were assigned. At that meeting, the benchmark was defined as '*a protocol to obtain a measure of performance of control strategies for activated sludge plants based on numerical, realistic simulations of the controlled plant*'. At that time a number of strategic decisions were taken and most of those decisions have had a large impact on the final version of BSM1. It was decided that specific control strategies should not be included in the benchmark definition. Moreover, the benchmark should not simply be simulation code, but also should include definitions of influent disturbances and performance evaluation criteria. It was decided that the first layout would be aimed at carbon/nitrogen removal in a series of tanks. Discussion groups were set up to decide upon the final clarifier model, the influent description and the performance assessment procedure. The full set of equations, all the parameter values, the procedure to be applied to verify the code and simulation results were all made available for use and discussion on the dedicated website http://www.benchmarkwwtp.org/ organised within the COST Action 682. The benchmarkers exchanged results electronically with Dr. Ulf Jeppsson from Lund University (Sweden) who was in charge of coordinating the information.

A subsequent meeting was held in Ljubljana (Slovenia) in April 1998. The main issues discussed in Slovenia were the size of the clarifier, the initialisation procedure, the influent description and the performance assessment. Implementation of the benchmark was discussed in Lund (Sweden) in August 1998. At that time, the benchmark had been implemented on six platforms: three implementations were coded from scratch (FORTRAN, C and Matlab/Simulink) and three were implemented in commercial WWTP software packages (GPS-X, Simba and WEST). Ongoing ring-tests revealed large discrepancies in computation time. Furthermore, these ring-tests revealed errors in the implementation code (in all platforms) and potential problems related to algebraic loops (due to the recycle loops in the plant). This ring-testing exercise also identified the impact of subtle differences in the implementation of the models. That is, not all the models were input into the commercial simulators in exactly the same way (aeration, soluble components in the settler,...) and the ring-testing identified the impact of these differences. Implementing code from scratch allowed more flexibility as the written equations were translated precisely into code, but the disadvantage was that the implementation was more prone to errors.

The meeting in Sweden was the last one under the umbrella of COST Action 682 and so the benchmark development migrated to COST Action 624 'Optimal Management of Wastewater Systems'. A meeting in Grenoble (France) (March 1999) was devoted to refinements of the basic benchmark. All the details were made available on the website and were published in a 'benchmark manual' (Copp, 2002) documenting BSM1. When preparing the document, a substantial effort went into verifying the steady state and dynamic output data included in the description. The manual summarises the various tested implementations with helpful hints for new benchmark users. Results were verified using BioWin, EFOR, GPS-X, Matlab/Simulink, Simba, STOAT, WEST and a user-defined FORTRAN code (Alex *et al.* 1999; Pons *et al.* 1999; Copp, 2002).

When COST Action 624 came to an end in 2004, the BSM1 development continued as the IWA Task Group on Benchmarking of Control Strategies for Wastewater Treatment Plants. At this point, the main focus was the definition of an updated set of evaluation criteria and a set of standardised sensor models. Some adjustments to BSM1 were also made to improve its compatibility with BSM1_LT and BSM2 (Sections 2.2 and 2.3, respectively), both of which were subsequently developed by this Task Group.

During the entire BSM1 development period, a number of MSc and PhD studies benefited from its existence and numerous scientific papers have been published in which the simulation protocol or the tools that had been produced as a result are a central ingredient (Technical Report No. 15).

2.2 BENCHMARK SIMULATION MODEL NO. 1 LONG-TERM

Although a very flexible tool, BSM1 is not intended for long-term evaluations. In the BSM1 definition, the temperature is constant and the evaluation period is limited to one week for three different weather scenarios (Copp, 2002). However, it was realised that many of the control actions at a WWTP have an effect on the process over longer time scales: sludge age control, equalisation basin control and sludge storage to mention a few. Also, BSM1 is mainly based on ideal sensors and actuators, because short-term evaluations do not allow for realistic equipment (sensor/actuator) modelling, including failures, drift and maintenance, as these typically appear over longer time scales than one week.

The Benchmark Simulation Model No. 1 Long-Term (BSM1_LT) addresses these BSM1 shortcomings and BSM1_LT also allows for process monitoring. Process monitoring is closely linked to process control. Process monitoring in this context means tracking measurement variables to detect process deviations, failures and faults. Process monitoring also involves the isolation of variables contributing to the deviations, facilitating further analysis of the problems. Monitoring of wastewater treatment operations has in the last decade become an intensive area of research and many different methods have been proposed (Rosen *et al.* 2003). Unfortunately, before the development of BSM1_LT there was no objective way to compare the success of methods for wastewater treatment monitoring, as researchers generally have used real data specific for a certain plant and, thus, not generally available to others. The BSM1_LT is intended to resolve this problem and is the first simulation platform that allows the objective comparison of WWTP monitoring methods.

The basic BSM1_LT concepts were discussed in detail during a COST Action 624 meeting in Lund (Sweden) in December 2003, and were then presented at the AutMoNet 2004 conference in Vienna (Austria). Those basic BSM1_LT concepts were published in Rosen *et al.* (2004), and have been developed further under the umbrella of this Task Group. First, a long-term influent file consistent with the previously developed BSM2-influent was created (Chapter 4). Existing BSM1 models were extended to include temperature dependency (see also Chapter 4). Sensor models that adopted the principles and models described by Rieger *et al.* (2003) were implemented. Models for sensor and actuator faults (Rosen *et al.* 2008) were also developed and incorporated. BSM1_LT required novel and suitable performance evaluation criteria (Corominas *et al.* 2011), and this proved to be one of the most difficult tasks to accomplish during the development of BSM1_LT. Ring-tested implementations of the BSM1_LT are available in Matlab/Simulink and WEST.

2.3 BENCHMARK SIMULATION MODEL NO. 2

A further limitation of BSM1 is that it does not allow for evaluation of control strategies on a plant-wide basis. BSM1 only includes an activated sludge system and a secondary clarifier. Consequently, only local control strategies can be evaluated. During the last decade the importance of integrated and plant-wide control has been emphasised by the research community and the wastewater industry is starting to realise the benefits of such an approach. A WWTP should be considered a unit, where primary/secondary clarification units, activated sludge reactors, anaerobic digesters, thickeners, dewatering systems and so on are linked together and need to be operated and controlled not only on a local level as individual processes but also by supervisory systems taking into account all the interactions between the processes. Otherwise, sub-optimal performance will be an unavoidable outcome leading to reduced effluent quality and/or higher operational costs.

It was the intent of the Benchmark Simulation Model No. 2 (BSM2) to take these issues into account. BSM1 was updated by adding wastewater pre-treatment and a sludge train including anaerobic digestion (Chapter 3). To allow for a more thorough and extended evaluation, the benchmark evaluation period was extended to one year (compared to one week in BSM1). The slow dynamics of anaerobic digestion processes also necessitated a prolonged evaluation period. With this extended evaluation period, it was reasonable to include seasonal effects on the WWTP in terms of temperature variations. The influent data files included with BSM1 (dry, storm and rain weather data) have been used extensively by researchers. However, for BSM2 the extended dynamic influent is generated with a phenomenological influent model (Gernaey *et al.* 2011).

The BSM2 concepts were first discussed in detail during a COST Action 624 meeting in Lund (Sweden) in December 2003 and were then presented at WaterMatex2004 in Beijing (China) and subsequently published by Jeppsson *et al.* (2006). Following the Lund meeting in 2003, the BSM2 development was coordinated by Dr. Ulf Jeppsson (Lund University, Sweden). Development continued under the umbrella of the Task Group starting in May 2005. The first BSM2 layout was based on the ideas outlined in Jeppsson *et al.* (2006). It was implemented and verified in accordance with the plan and presented at WaterMatex2007 in Washington D.C. (USA) (Jeppsson *et al.* 2007). Major milestones in the BSM2 development were the implementation and verification of the Anaerobic Digestion Model No. 1 (ADM1, Batstone *et al.* 2002; Rosen *et al.* 2006), the development of phenomenological influent models (Gernaey *et al.* 2011) and the development of ASM1 → ADM1 and ADM1 → ASM1 model interfaces (Nopens *et al.* 2009).

The BSM2 layout presented at WaterMatex2007 included 15 simple demonstration cases, both with and without active controllers, and was aimed at investigating how the evaluation criteria captured various operational conditions. It was however revealed that: (1) the evaluation criteria were not very sensitive to the different cases tested; and, (2) the very highly loaded system, which was deliberately adopted, limited what could be accomplished by active control. Indeed, it was shown that active control has its limitations and will not be able to significantly improve the performance of a highly overloaded plant. Interestingly, the high nitrogen load which was causing some of the issues was associated with the reject water, which was not present or accounted for in the BSM1 case. The final BSM2 layout was presented by Nopens *et al.* (2010). This layout (compared to the earlier versions) included: (1) a reduced N load to compensate for the contribution of the reject water; (2) increased activated sludge tank volumes, in order to obtain a low/medium loaded WWTP that can benefit from process control; and, (3) modifications of the evaluation criteria. The new criteria made a distinction between nitrate and ammonium nitrogen and reduced the dominating effect of aeration that had been observed in BSM1 and earlier versions of the BSM2. Ring-tested implementations of the BSM2 are available in Fortran, Matlab/Simulink, Simba and WEST.

2.4 THE BENCHMARK SIMULATION MODEL SET

From the outset, the BSM1, BSM1_LT and BSM2 protocols have been developed to help with the unbiased evaluation of control-related simulations. Through the years, issues have been identified and short-comings in the models have been debated. The current protocols represent the state-of-the-art and encompass a scope meant to be representative of most wastewater treatment control applications. However, with each new application, and there have been hundreds, comes suggestions for improvements and although this STR deals only with these three protocols, further developments are fully expected and encouraged in the future.

Chapter 3

Benchmark plant description

U. Jeppsson, J. B. Copp, K. V. Gernaey, M.-N. Pons and P. A. Vanrolleghem

This chapter of the STR provides a brief overview description of the different WWTP simulation benchmarks that are distributed by the Task Group. In practice, this chapter provides a description of the benchmark-specified equipment in each benchmark plant. Furthermore, two reference configurations have been defined for each benchmark plant. These reference configurations are referred to as the 'open-loop' and 'closed-loop' configurations. The 'open-loop' configuration (without any control) provides a reference point to compare the performance of a control and/or monitoring strategy and the 'closed-loop' configuration is the same configuration with a defined and active default control strategy added to it. This chapter describes the open-loop reference configuration. A full description of the models, evaluation criteria and simulation procedures are described in subsequent chapters. The closed-loop configurations (i.e., the default control strategies) are described in Chapter 5.

3.1 BENCHMARK SIMULATION MODEL NO. 1

The purpose of the BSM1 system was to define a tool that would allow for the objective and unbiased comparison of carbon and nitrogen removal activated sludge control strategies. The BSM1 plant is a simple layout consisting of an activated sludge tank and a secondary clarifier as schematically shown in Figure 3.1. For the open-loop configuration, both liquid flow rates and aeration intensity are fixed.

The physical attributes of the open-loop BSM1 layout (Figure 3.1) are defined as follows:

- The temperature of the influent and the mixed liquor is constant at 15°C
- The average dry weather influent flow rate to the plant is 18 446 $m^3 \cdot d^{-1}$
    - step feed of the influent is possible, but the open-loop configuration has all of the influent flow directed into the first unaerated tank
    - influent composition is described in Chapter 4
- The activated sludge tank consists of 5 tanks-in-series, with a total biological volume of 5 999 m^3
    - tanks 1 and 2 are 1 000 m^3 each, tanks 3, 4 and 5 are 1 333 m^3 each
    - tanks 1 and 2 are unaerated in open-loop, but fully mixed

  - tanks 3, 4 and 5 are aerated
  - for the open-loop case, the oxygen transfer coefficients (K_La) are fixed
  - for tanks 3 and 4 the coefficient is set to 240 d^{-1} (10 h^{-1})
  - for tank 5 the coefficient is set to 84 d^{-1} (3.5 h^{-1})
  - the DO saturation concentration in tanks 3, 4 and 5 is set to 8 g $O_2 \cdot m^{-3}$
- The secondary clarifier has a volume of 6 000 m^3 (area of 1 500 m^2 and a depth of 4 m)
  - the feed point to the settler is at 2.2 m from the bottom.
- Waste activated sludge (WAS) is pumped continuously from the secondary settler underflow at a default flow rate of 385 $m^3 \cdot d^{-1}$
- There are two recycle flows
  - a 55 338 $m^3 \cdot d^{-1}$ internal recycle from the 5th to the 1st tank
  - a 18 446 $m^3 \cdot d^{-1}$ RAS recycle from the secondary settler underflow to the 1st tank.

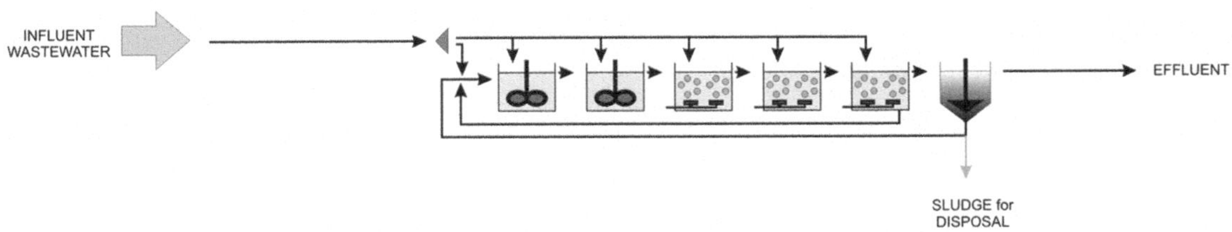

Figure 3.1 Schematic representation of the *BSM1* plant design.

3.2 BENCHMARK SIMULATION MODEL NO. 1 LONG-TERM

The basic features of the BSM1_LT plant layout are identical to BSM1. The differences between BSM1_LT and BSM1 are:

- The evaluation period is extended to 52 weeks (1 year)
- Long-term dynamics are introduced
  - the extended influent includes dry weather as well as rain events, [i.e., the BSM1 approach of using separate 14-day influent files (one for dry weather, one for rain, and one for a storm event) was altered so that one influent file included all of these types of events over an extended period]
  - the average influent flow rate for the evaluation period is 20 504 $m^3 \cdot d^{-1}$
  - the influent and mixed liquor temperatures vary with time
  - details on the influent dynamics and composition are provided in Chapter 4
- Sensor, actuator and process faults are included and described in Chapter 4.

3.3 BENCHMARK SIMULATION MODEL NO. 2

The main driver for the development of BSM2 was to extend the BSM concept to a long-term and plant-wide context. Following the development and subsequent experience with BSM1, it was realised that the lack of sludge handling in BSM1 could limit and bias the comparison of different control strategies, even though things like sludge production were being considered in the evaluation criteria. The BSM2 concept was developed to address these perceived short-comings. BSM2 (Figure 3.2) is based on BSM1 for the biological treatment, but, BSM2 includes a primary clarifier, a thickener for the sludge wasted from the BSM1 secondary clarifier, an anaerobic digester for treatment of thickened secondary sludge and solids wasted from the primary clarifier and a dewatering unit for the digested sludge. The effluent liquid streams from the thickening and dewatering steps are recycled back to the inlet of the primary clarifier. With these additions, different control handles such as pumps and valves have been included (Figure 3.2).

The details of the open-loop plant are as follows:

- The influent and mixed liquor temperatures vary with time
- The average influent flow rate for the evaluation period is 20 648 $m^3 \cdot d^{-1}$
  - similar to BSM1_LT, this influent includes dry weather as well as rain and storm events and covers an extended period of time (1 year)
  - details of the influent dynamics and composition are provided in Chapter 4
- The primary clarifier has a volume of 900 m^3 (area of 300 m^2, depth of 3 m)
  - the default underflow from the primary clarifier is 147.6 $m^3 \cdot d^{-1}$

NOTE: While developing BSM2, it was discovered that the BSM1 biological tanks were severely overloaded with nitrogen partly as a consequence of the nitrogen rich reject water recycled from the dewatering step. As a result, for BSM2, the influent nitrogen load was reduced and tank volumes were re-evaluated on the basis of the design guidelines from both the German Association for Water Economy, Wastewater and Waste (ATV A131, 2000) and the US Environmental Protection Agency (Harris *et al.* 1982), as explained in Nopens *et al.* (2010). Both guidelines suggested that the volume of the activated sludge tanks be increased considerably.

- The activated sludge tank is configured as five (5) tanks-in-series with a total volume of 12 000 m^3
  - the first two unaerated tanks have a volume of 1 500 m^3 each
  - the three aerobic tanks have a volume of 3 000 m^3 each
  - the default K_La coefficients in the aerobic tanks are fixed, 120 d^{-1} (5 h^{-1}) for tanks 3 and 4 and 60 d^{-1} (2.5 h^{-1}) for tank 5
- Two activated sludge recycles are included
  - a 20 648 $m^3 \cdot d^{-1}$ RAS recycle from the secondary settler underflow to the 1st tank
  - an internal recycle with a default flow rate of 61 944 $m^3 \cdot d^{-1}$ from the 5th to the 1st tank
- Supplemental carbon is dosed to the first anoxic tank, at a default flow rate of 2 $m^3 \cdot d^{-1}$ and concentration of 400 kg COD $\cdot$ m^{-3}
- The secondary clarifier has a volume of 6 000 m^3 (area of 1 500 m^2, depth of 4 m)
  - the feed point to the clarifier is at 2.2 m from the bottom
- The anaerobic digester (AD) has a liquid volume of 3 400 m^3 and gas volume (headspace) of 300 m^3
- The default WAS flow rate is 300 $m^3 \cdot d^{-1}$ and is fed into the thickener
- The thickener and the dewatering units have no volume
  - these units are considered to be ideal point separators that, at steady state, produce concentrated sludge flows of 7% and 28% TSS, respectively
  - the underflow from the thickener is 30.9 $m^3 \cdot d^{-1}$
  - the dewatering generates 9.6 $m^3 \cdot d^{-1}$ of concentrated sludge and 168.9 $m^3 \cdot d^{-1}$ of reject water
- The reject water storage tank has a volume of 160 m^3
  - in the open-loop case the storage tank is not used, that is, the reject water bypasses the storage tank and is combined with the influent before the inlet of the primary clarifier.

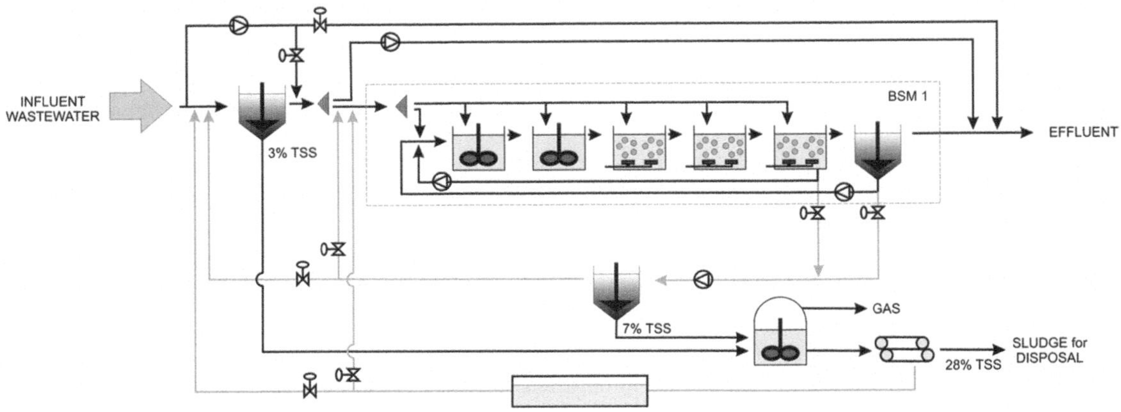

Figure 3.2 Schematic representation of the *BSM2* plant design. In addition to the BSM1 unit processes, BSM2 includes a primary clarifier (PRIM), a thickener (THK), an anaerobic digester (AD), a dewatering unit (DW) and a storage tank (ST) for the reject water originating from the DW.

3.4 CHARACTERISTICS SUMMARY

The following tables summarise the key physical and operational characteristics of the benchmark models. The physical attributes of the biological reactors and the settlers are listed in Table 3.1 and some of the system variables for each configuration are listed in Table 3.2.

Table 3.1 Physical attributes of the biological reactors and settling tank for the *simulation benchmark* process configurations.

	Physical configuration			
	BSM1	**BSM1_LT**	**BSM2**	**Units**
Depth – primary settler	n/a	n/a	3	m
Area – primary settler	n/a	n/a	300	m^2
Volume – primary settler	n/a	n/a	900	m^3
Volume – tank 1	1 000	1 000	1 500	m^3
Volume – tank 2	1 000	1 000	1 500	m^3
Volume – tank 3	1 333	1 333	3 000	m^3
Volume – tank 4	1 333	1 333	3 000	m^3
Volume – tank 5	1 333	1 333	3 000	m^3
Depth – secondary clarifier	4	4	4	m
Area – secondary clarifier	1 500	1 500	1 500	m^2
Volume – secondary clarifier	6 000	6 000	6 000	m^3
Volume – anaerobic digester	n/a	n/a	3 400	m^3
Volume – digester head space	n/a	n/a	300	m^3
Volume – reject water storage	n/a	n/a	160	m^3

Table 3.2 A selection of system variables.

	Default system flows			Units
	BSM1	**BSM1_LT**	**BSM2**	
Average influent flow	18 446	20 504	20 648	$m^3 \cdot d^{-1}$
Recycle flow	18 446	18 446	20 648	$m^3 \cdot d^{-1}$
Internal recycle flow	55 338	55 338	61 944	$m^3 \cdot d^{-1}$
K_La – Tank 3	240	240	120	d^{-1}
K_La – Tank 4	240	240	120	d^{-1}
K_La – Tank 5	84	84	60	d^{-1}
Wastage flow	385	385	300*	$m^3 \cdot d^{-1}$
Primary clarifier underflow	n/a	n/a	147.6	$m^3 \cdot d^{-1}$
Supplemental carbon conc.	n/a	n/a	400	kg COD $\cdot m^{-3}$
Supplemental carbon flow	n/a	n/a	2	$m^3 \cdot d^{-1}$
Thickener underflow conc.	n/a	n/a	7	% TSS
Thickener underflow	n/a	n/a	30.9	$m^3 \cdot d^{-1}$
Dewatering sludge conc.	n/a	n/a	28	% TSS
Dewatering sludge flow	n/a	n/a	9.6	$m^3 \cdot d^{-1}$
Dewatering reject water flow	n/a	n/a	168.9	$m^3 \cdot d^{-1}$

*During cold season (<15°C).

Chapter 4

Benchmark models

J. Alex, D. Batstone, L. Benedetti, J. Comas, J. B. Copp, L. Corominas, X. Flores-Alsina, K. V. Gernaey, U. Jeppsson, I. Nopens, M.-N. Pons, I. Rodríguez-Roda, C. Rosen, J.-P. Steyer, P. A. Vanrolleghem, E. I. P. Volcke and D. Vrecko

The benchmark simulation models (BSMs) are the result of the combination of a number of sub-models that have been used to mathematically describe each process in the different benchmark layouts (Chapter 3). This chapter provides a brief description of those benchmark sub-models. Some of these models are specific to particular unit processes (i.e., ASM1) and others are globally applicable (i.e., temperature).

The first sections of this chapter address a new influent wastewater generation model specifically developed for the BSMs, but universally applicable. This is followed by sections on individual unit process models, sensors and actuators, and inhibition and toxicity modelling. The final sections address more general concepts including a new approach aimed at predicting the possibility of settling problems and finally the methods used for modelling temperature and the effect of temperature on kinetic parameters.

Table 4.1 shows the various unit process models, and the benchmark that they are associated with.

Table 4.1 Modelling capability included in the three BSMs. A '✓' signifies that the benchmark includes that behaviour or capability, but its implementation is simulation-specific.

	Benchmark model		
	BSM1	**BSM1_LT**	**BSM2**
Influent	14-day data files	model generated 609-day data file	model generated 609-day data file
Activated sludge	ASM1	ASM1	ASM1
Anaerobic digester	–	–	ADM1
Primary clarifier	–	–	Otterpohl

(*Continued*)

Table 4.1 Modelling capability included in the three BSMs. A '✓' signifies that the benchmark includes that behaviour or capability, but its implementation is simulation-specific (*Continued*).

	Benchmark model		
	BSM1	**BSM1_LT**	**BSM2**
Secondary sedimentation	Takács	Takács	Takács
Thickener	–	–	ideal mass balance
Dewatering	–	–	ideal mass balance
Reject water storage	–	–	non-reactive tank
Sensors and actuators	✓	✓	✓
Inhibition and toxicity	✓	✓	✓
Temperature impacts	–	✓	✓

NOTE: Space limitation requires that all of the BSM sub-models be summarised only here. In this chapter the reader will find brief model overviews and discussions related to specific model features of interest (new findings, short-comings in the published models, . . .) related to the BSM implementations. This chapter introduces all of the models, but for detailed descriptions of the various models, the reader is referred to the original publications and the detailed technical reports that accompany this STR.

4.1 INFLUENT MODELLING

In any wastewater treatment modelling project, the influent is the single most important feature that needs to be well understood. This is equally true for the benchmark. The influent load impacts sludge production, fractionation of that load impacts the aeration requirements and nutrient removal possibilities and the dynamics in the influent drive the short- and long-term process dynamics. The influent disturbances used to test a particular control strategy play a critical role in the evaluation. That is, because of the multi-faceted nature of the activated sludge process, a particular control strategy may react well to one disturbance and not well to another. Hence, for an unbiased and complete evaluation, it is important that a series of disturbances be defined and that each control strategy be subjected to all the disturbances. Only then can a fair comparison be made.

The influent to the benchmark models has undergone several iterations over the years. BSM1 used a series of 14-day data files, but this approach was insufficient for BSM1_LT and BSM2, so a comprehensive influent model was developed. This section provides only an overview of the different influent options, so the reader is referred to the detailed technical reports or published literature for a complete and detailed description of the approaches.

4.1.1 BSM1 influent

The influent to BSM1 is based on three different 'weather files', corresponding to dry, storm and rain weather disturbance scenarios (Vanhooren & Nguyen, 1996; Spanjers *et al.* 1998; Copp, 1999; Copp, 2002; Copp *et al.* 2002). Each of the files contains 14 days of influent data at 15-minute intervals and defines the influent – in this case, the influent is assumed to be primary effluent – using the ASM1 state variables (Section 4.2.1). In general, these files were developed based on real data and depict expected diurnal variations in influent flow, as well as COD and nitrogen concentrations. The *dry weather* file depicts what is considered to be *normal* diurnal variations. In this file, the resultant peaking factor is 1.74 for maximum flow and 2.34 for maximum COD mass load (i.e., flow∗ concentration, mass/day) as compared to the flow-weighted average values. The second file is a variation on the first with the incorporation of two storm events in the second week. The first storm event in this file is a high intensity and short duration event and is expected to 'flush' the sewer of particulate material. The resuspension of these particles is reflected in the data through a significant increase in inert and biodegradable suspended solids. The second storm event assumes the sewers were cleared of particulate matter during the first storm event; hence, only a modest increase in COD load is noted during the second storm. This behaviour is assumed even though the peak flow for both storms is the same and the peak flow of the second storm is maintained over a longer period of time. The third file is meant to represent a long rain event in the second week. The influent flow during this rain event does not reach the level attained during the storm events, but the increased flow is sustained for a much longer period of time. Unlike the storm events, there is no increase in COD load to the plant during the rain event. The flow-weighted average concentrations of the influent components for the three files are shown in Table 4.2.

Table 4.2 Flow-weighted average influent composition in the influent files.

Component	Dry weather	Storm event	Rain event	Units
S_S	69.50	64.93	60.13	$g\ COD \cdot m^{-3}$
$X_{B,H}$	28.17	27.25	24.37	$g\ COD \cdot m^{-3}$
X_S	202.32	193.32	175.05	$g\ COD \cdot m^{-3}$
X_I	51.20	51.92	44.30	$g\ COD \cdot m^{-3}$
S_{NH}	31.56	29.48	27.30	$g\ N \cdot m^{-3}$
S_I	30.00	28.03	25.96	$g\ N \cdot m^{-3}$
S_{ND}	6.95	6.49	6.01	$g\ N \cdot m^{-3}$
X_{ND}	10.59	10.24	9.16	$g\ N \cdot m^{-3}$
Q	18 446	19 745	21 320	$m^3 \cdot day^{-1}$

4.1.2 BSM1_LT and BSM2 influent

When the primary clarifier was added to the definition of Benchmark Simulation Model No. 2 (BSM2), it was assumed that the influent to this settler could be adjusted to achieve a primary effluent that was identical to the BSM1 influent. Unfortunately, this assumption proved to be incorrect as it was not possible to achieve the exact BSM1 influent dynamics (equivalent to the BSM2 primary effluent) with a reasonable raw influent to the primary clarifier model. As a result, a separate approach to generate dynamic influent disturbances was adopted for BSM2 and this approach was subsequently used for BSM1_LT as well.

The development of BSM1_LT and BSM2 also highlighted another issue. The evaluation procedure in BSM1 is based on seven (7) days of output data so the use of 28-day influent files was deemed sufficient (two 14-day files back-to-back). However, not all control options can be suitably evaluated within a 7-day period. This is especially true for strategies that involve 'slow' process variables such as solids retention times and/or waste sludge rates. To address this limitation for BSM1_LT and BSM2, the simulation period was increased from 28 days to 18 months (with the last twelve months used for the evaluation) and a new influent modelling tool (Gernaey *et al.* 2006, 2011) was developed to generate the influent for these simulations. This new tool was used to generate influent data for both BSM1_LT and BSM2 even though BSM2 includes a primary clarifier, and BSM1_LT does not. The generic nature of the tool enabled the generation of a primary effluent influent for BSM1_LT and a raw wastewater influent for BSM2.

Lastly, even though the influents for BSM1_LT and BSM2 are fully defined, it must be understood that the data are based on a model and that the influent model includes typical phenomena that are observed in a year of full-scale WWTP influent data:

- Diurnal behaviour;
- Weekend pattern, which consists of a lower average flow rate and lower pollutant loads during weekends compared to normal weekdays;
- Seasonal changes to model increased infiltration in the rainy season compared to the dry season;
- Holiday periods during which time lower average wastewater flow rates are maintained over a period of several weeks;
- Temperature dynamics;
- Rain events.

The complete model is based on a series of sub-models that, when combined, produce an influent that reflects the various phenomena discussed above. The dry weather model for instance might be combined with rain and storm weather generation to account for 'first flush' effects from the sewer network and dilution phenomena that are typically observed at full-scale WWTPs. An industrial block is included to account for industrial loads and a temperature model is included so that temperature effects also can be described.

Flow rate design principles

The model blocks used to generate the influent flow rate profiles are shown in Figure 4.1. The BSM1 dry weather influent ($Q_{av} = 18\,446\ m^3 \cdot d^{-1}$; Q_{av} on weekdays = $19\,346.3\ m^3 \cdot d^{-1}$; Q_{av} on weekend days = $16\,196.3\ m^3 \cdot d^{-1}$) was used as the basis for the generation of the new influent flow profile. Using these BSM1 flows, assumptions were made with respect to the flow contributions from infiltration, households and industrial sources (~25%, ~62% and ~13%, respectively).

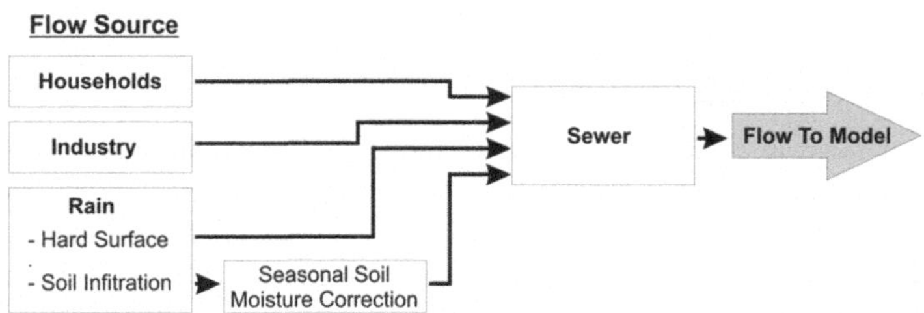

Figure 4.1 General overview of the influent flow rate model.

As illustrated in Figure 4.1, the generation of the influent flow rate is achieved by combining the contributions from households, industry, infiltration and rain with the combination giving the overall sewer flow rate. This flow is then passed through a simple sewer system model (variable volume reservoir series) to achieve the final benchmark influent flow. Each model block that makes up Figure 4.1 is the combination of a number of underlying models and assumptions. As an example, details of the *Households* model block are shown in Figure 4.2.

The basic steps in the flow generation from the *Households* model are described here, but the reader is referred to the detailed technical report and published articles for a more complete description.

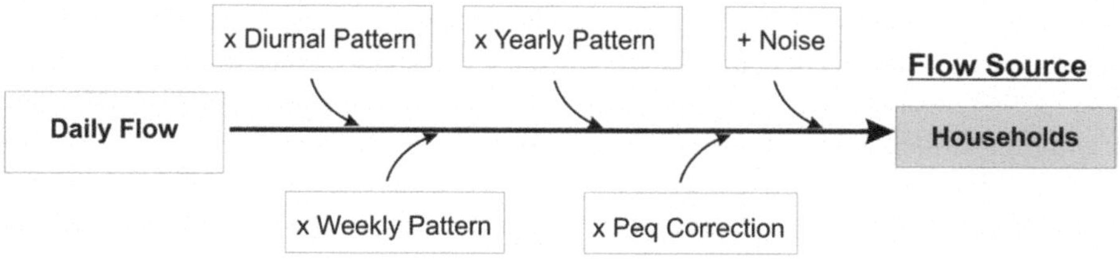

Figure 4.2 *Households* model block, with diurnal flow rate variations, weekend effects and holiday effects.

The basic approach involves a diurnal profile corresponding to the diurnal variation of the wastewater production of one person equivalent (peq) that is normalised and then multiplied by a daily flow rate per person (i.e., 150 $(L \cdot d^{-1}) \cdot peq^{-1}$) to give a dynamic daily flow profile per person. A second function is used to superimpose a weekly household flow rate pattern on the diurnal profiles. In the BSM case, the weekend effect is a slight reduction in the household wastewater production (8% reduction on Saturday, 12% on Sunday) compared to normal weekdays. Similar to the weekend effect, a holiday effect that decreases the flow for three (3) weeks, is added on top. The holiday effect results in a 25% decrease in the flow during the first two weeks, and a 12% decrease during the third holiday week. To increase the realism in the data, white noise, generated when a random number is multiplied by a gain, is added to the calculated flow rate. A sanity check is included to avoid negative values, and to avoid unrealistically high flow rates. Finally, the wastewater production is multiplied by the number of person equivalents in the catchment area to give the estimated household contribution to the total influent flow.

Figure 4.3 illustrates the impact of the various blocks on the calculated diurnal flow rate profiles generated by the *Households* model. Figures 4.3a and 4.3b show the impact of adding noise. Figure 4.3c indicates the impact of the weekend effect and Figure 4.3d shows the effect of adding noise. The holiday effect is presented in Figures 4.3e and 4.3f. These figures show only the average daily flows to minimise the number of data points, but the figures clearly demonstrate a significant reduction in the influent flow rate for a number of days. When tested (without noise), this approach produced an annual average household flow rate equal to 11 513 $m^3 \cdot d^{-1}$, which is very close to 62% of 18 446 $m^3 \cdot d^{-1}$ which was the assumed household contribution in the BSM1 influent flow.

A similar approach is used to calculate the industrial contribution to the influent flow. A weekly pattern including a Friday and weekend effect is combined with a holiday effect that corresponds to the same period as the *Household* holiday period. The *Industry* contribution also includes a one-week holiday effect towards the end of the year. Noise is added and the final result is passed through a sanity check to avoid negative flow rates or rates that are too high.

Rainfall is assumed to contribute to the total flow in two ways: the largest portion of the rainfall flow is assumed to be runoff from impervious surfaces. However, it is also assumed that rainfall on pervious surfaces will influence the groundwater level and contribute to infiltration. The model assumes two seasons and the *Seasonal correction infiltration* model is used to create a dry and wet weather seasonal effect. This seasonal correction is combined with the rainfall and the net contribution of infiltration is calculated. The details related to the calculation of the rainfall portion of the influent flow are described elsewhere (Gernaey *et al.* 2011).

Composition design principles

Similar to the influent flow generation, it is assumed that there are two pollutant sources: households (*Households pollutants* model) and industry (*Industry pollutants* model). The complexity of the model is reduced significantly by neglecting all other influent pollutant sources (infiltration and rain). Pollutant fluxes from both sources are calculated and combined and then the particulate concentrations are altered based on flow to account for first flush effects. The resulting influent vector, consisting of the 13 ASM1 state variables, the TSS concentration (as defined in Chapter 6), a flow rate and temperature, can be used directly as an input to any ASM1 compatible simulation model. The BSM pollutant loads were designed based on the assumption that 20% of the COD load and 10% of the nitrogen load are due to industrial activity in the BSM1 influent.

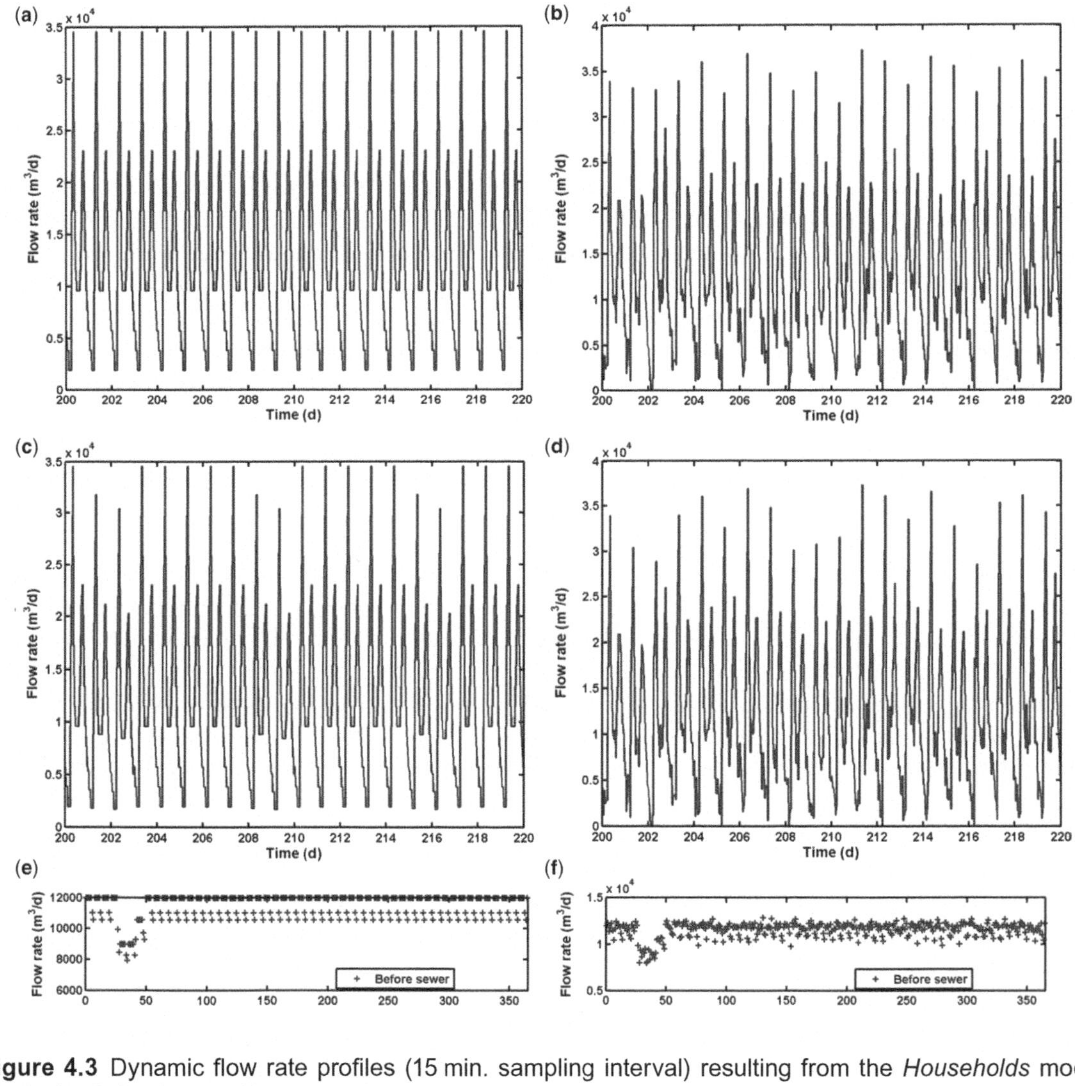

Figure 4.3 Dynamic flow rate profiles (15 min. sampling interval) resulting from the *Households* model block: (a) Only diurnal effect, no noise added; (b) As (a), but with noise added in the *Households* model block; (c) Diurnal effect combined with weekend effect, no noise added; (d) As (c), but with noise added; (e) Daily average flow rates (each average daily flow rate value represents the mean of 96 samples), combining diurnal, weekend and holiday effects, no noise added; (f) As (e), but with noise added. Note that the first day in (e) and (f) corresponds to July 1st.

A generalised approach, based on particulate COD (COD_{part}), soluble COD (COD_{sol}), ammonium (S_{NH}) and Kjeldahl nitrogen (TKN) has been adopted for the composition of the influent (Figure 4.4). The advantage of this approach is that these components can be used for generating influent files for other ASM models making this tool universally applicable and not simply a benchmark model (Gernaey *et al.* 2011).

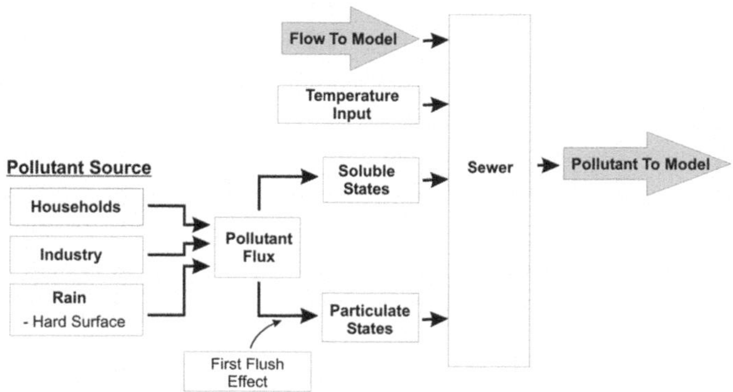

Figure 4.4 Overview of the modules in the proposed model for influent data generation.

Diurnal profiles (24 values per day) for soluble and particulate COD (COD_{sol}, COD_{part}), ammonium (S_{NH}) and Kjeldahl nitrogen (TKN) fluxes form the basis for the dynamic profiles generated with the *Households pollutants* model (Figure 4.5). The values in the four input files are normalised and the pollutant fluxes are transformed into (g COD$\cdot$peq^{-1})$\cdot$d^{-1} and (g N$\cdot$peq^{-1})$\cdot$d^{-1} units by multiplying the values in the input files by specific scaling constants (i.e., a different constant for each profile) and then by a constant representing the number of person equivalents in the catchment. These mass load profiles are then subjected to transformations that account for the weekend and holiday effects. The weekend effect corresponds to a reduction of the household pollutant fluxes (12% reduction on Saturday, 16% on Sunday) and the holiday effect uses the same reduction approach as for the flow rates. Uncorrelated white noise is added to each pollutant flux and a sanity check is applied to avoid negative values.

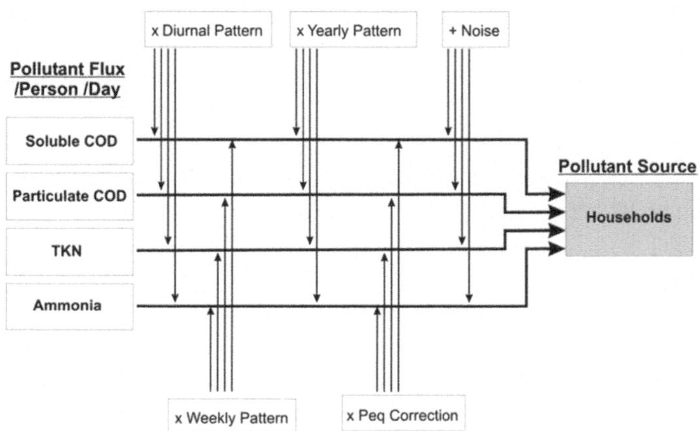

Figure 4.5 Layout of the *Households pollutants* model.

The *Industry pollutants* model (Figure 4.6) has a similar structure. Weekly profiles for COD_{sol}, COD_{part}, S_{NH} and TKN that contain diurnal pollutant flux variations, a Friday afternoon pollutant flux peak load and a weekend effect form the basis for the industrial pollutant fluxes. A holiday effect is added using the same input approach as for the industrial flow. After multiplying the weekly flux pattern with the holiday pattern, zero mean white noise is added and a sanity check is applied to avoid negative values.

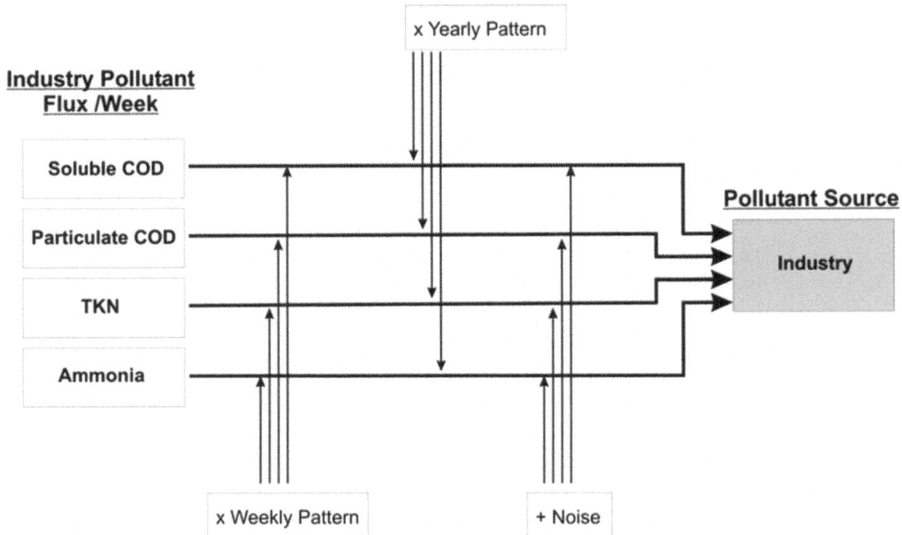

Figure 4.6 Layout of the *Industry pollutants* model.

With the loads determined, an *ASM1_fractionator* model converts the bulk pollutant concentrations to state variable concentrations that are compatible with the ASM1 model (Table 4.3). Several basic principles are used to generate the state variables. For instance, the following rules were used:

- Provided that the COD_{sol} flux is greater than 30 g $COD \cdot m^{-3}$, the S_I concentration is assumed to be constant and equal to 30 g $COD \cdot m^{-3}$. It is furthermore assumed that S_I is present in the dry weather wastewater flow. Only rainfall runoff from impervious areas can dilute S_I.
- The S_S concentration is calculated as the difference between the COD_{sol} flux and the S_I pollutant flux.
- $X_{B,A}$ and X_P are assumed to be zero.
- The COD_{part} is distributed between X_I, X_S and $X_{B,H}$ according to the BSM1 flow weighted average dry weather influent (18.2%, 71.8% and 10.0%, respectively).
- The S_O (oxygen) is assumed to be zero.
- The S_{NO} (nitrate) is zero.
- The S_{NH} concentration is calculated based on the S_{NH} flux.
- The organic N flux is obtained by subtracting the S_{NH} flux and the N flux associated with the organic particulate variables, $X_{B,H}$ and X_I. If the result of this calculation is greater than zero, the remaining organic N is distributed between S_{ND} and X_{ND} based on the S_{ND}/X_{ND} ratio in the BSM1 influent files (39.6% and 60.4%, respectively). If the result of this calculation is negative, influent S_{ND} and X_{ND} are assumed to be zero.

Table 4.3 Average flow-weighted influent concentrations calculated for one year of influent data generated with the BSM2 influent model.

Variable	Concentration	Unit
$S_{I,inf}$	27.21	$g\ COD \cdot m^{-3}$
$S_{S,inf}$	58.15	$g\ COD \cdot m^{-3}$
$X_{I,inf}$	92.46	$g\ COD \cdot m^{-3}$
$X_{S,inf}$	363.77	$g\ COD \cdot m^{-3}$
$X_{B,H,inf}$	50.66	$g\ COD \cdot m^{-3}$
$X_{B,A,inf}$	0.00	$g\ COD \cdot m^{-3}$
$X_{P,inf}$	0.00	$g\ COD \cdot m^{-3}$
$S_{O,inf}$	0.00	$g\ O_2 \cdot m^{-3}$
$S_{NO,inf}$	0.00	$g\ N \cdot m^{-3}$
$S_{NH,inf}$	23.85	$g\ N \cdot m^{-3}$
$S_{ND,inf}$	5.64	$g\ N \cdot m^{-3}$
$X_{ND,inf}$	16.12	$g\ N \cdot m^{-3}$
$S_{ALK,inf}$	7.00	$mol\ HCO_3^- \cdot m^{-3}$
TSS_{inf}	380.17	$g \cdot m^{-3}$
Q_{inf}	20 668.44	$m^3 \cdot d^{-1}$
T_{inf}	14.86	°C

4.2 UNIT PROCESS MODELS

To increase the acceptability of the results, four internationally accepted process models were chosen. Activated Sludge Model No. 1 (ASM1) was chosen as the biological process model (Henze *et al.* 1987), Anaerobic Digestion Model No. 1 (ADM1) was chosen as the anaerobic digestion model (Batstone *et al.* 2002), the Otterpohl (1995) model was chosen for primary settling and the double-exponential settling velocity function of Takács *et al.* (1991) was chosen as a fair representation of the secondary settling process.

4.2.1 Activated Sludge Model No. 1 (ASM1)

For the biological reactors, ASM1 was chosen for its universal appeal. It is acknowledged however, that there are several limitations with ASM1, and several activated sludge models have been developed since ASM1 was first introduced, including ASM2d and ASM3. Nevertheless, its practical validation and the international acceptance of ASM1 at the time the first benchmark was developed was the overriding criterion for its use. A complete description of the model and its development are available elsewhere (Henze *et al.* 1987). Table 4.4 lists the ASM1 state variables, the associated symbols and the state variable units.

To ensure the consistent application of the model in all benchmarking studies, all of the kinetic and stoichiometric model parameters have been defined in the simulation benchmark description (Appendix A). Included in these tables are the parameter descriptions, their recognised symbols and units as well as an associated value. The listed parameter estimates approximate those that are expected at 15°C.

Table 4.4 State variables for the IWA Activated Sludge Model No. 1 (ASM1).

State variable description	State symbol	Units
Soluble inert organic matter	S_I	$g\ COD \cdot m^{-3}$
Readily biodegradable substrate	S_S	$g\ COD \cdot m^{-3}$
Particulate inert organic matter	X_I	$g\ COD \cdot m^{-3}$
Slowly biodegradable substrate	X_S	$g\ COD \cdot m^{-3}$
Active heterotrophic biomass	$X_{B,H}$	$g\ COD \cdot m^{-3}$
Active autotrophic biomass	$X_{B,A}$	$g\ COD \cdot m^{-3}$
Particulate products arising from biomass decay	X_P	$g\ COD \cdot m^{-3}$
Dissolved oxygen	S_O	$g\ O_2 \cdot m^{-3}$
Nitrate and nitrite nitrogen	S_{NO}	$g\ N \cdot m^{-3}$
$NH_4^+ + NH_3$ nitrogen	S_{NH}	$g\ N \cdot m^{-3}$
Soluble biodegradable organic nitrogen	S_{ND}	$g\ N \cdot m^{-3}$
Particulate biodegradable organic nitrogen	X_{ND}	$g\ N \cdot m^{-3}$
Alkalinity	S_{ALK}	$mol\ HCO_3^- \cdot m^{-3}$

4.2.2 Anaerobic Digestion Model No. 1 (ADM1)

As with ASM1, the choice of ADM1 as the anaerobic digestion model was driven by its availability, potential for integration and international exposure. All of the models in the BSMs were thoroughly tested and this provided a great deal of insight into the model description as well as the model implementations in various platforms. The introduction of ADM1 into BSM2 was no exception. The BSM2 implementation identified several issues not explicitly discussed in the published model. The description here addresses some of those issues. It should be noted that the ADM1 implementation for BSM2 deviates somewhat from the model description in Batstone *et al.* (2002) and there are three main reasons for this. Firstly, ADM1 has been implemented so that it is consistent with the full BSM2. Secondly, computational issues have been addressed (Rosen *et al.* 2006) and lastly, unpublished parameter values have been chosen in consultation with the original ADM model developers (Batstone 2002–2008, personal communication). Readers are referred to the detailed technical report and original publication for a complete description of ADM1.

4.2.2.1 Elemental balances

Maintaining a complete elemental balance for all model components (COD, N, . . .) is a fundamental issue for any model and was explored in detail in Hauduc *et al.* (2010) where the authors attempted to verify several published activated sludge models. However, the implementation of ADM1 into the benchmark identified a few potential problems.

ADM1 includes a process referred to as disintegration, where composite material (X_c) is transformed into various compounds (soluble inerts (S_I), carbohydrates (X_{ch}), proteins (X_{pr}), lipids (X_{li}) and particulate inerts (X_I)) – all defined in COD units. Assuming one COD mass unit of X_c completely disintegrates, the published model assumes the following transformations:

$$\begin{cases} S_I &= f_{SI,xc} \cdot X_c = 0.1 \cdot X_c \\ X_I &= f_{XI,xc} \cdot X_c = 0.25 \cdot X_c \\ X_{ch} &= f_{ch,xc} \cdot X_c = 0.2 \cdot X_c \\ X_{pr} &= f_{pr,xc} \cdot X_c = 0.2 \cdot X_c \\ X_{li} &= f_{li,xc} \cdot X_c = 0.25 \cdot X_c \end{cases}$$

A COD balance exists as long as the sum of all $f_{i,xc} = 1$. However, this process introduces a potential nitrogen balance problem. The nitrogen content of X_c (N_{xc}) is 0.002 kmol N·(kg COD)$^{-1}$. If we calculate the nitrogen content of the disintegration products (kmol N·(kg COD)$^{-1}$) using parameter values from Batstone *et al.* (2002), we get:

$$\begin{aligned} N_{SI} \cdot 0.1 \cdot X_c + N_{XI} \cdot 0.25 \cdot X_c + N_{ch} \cdot 0.2 \cdot X_c + N_{aa} \cdot 0.2 \cdot X_c + N_{li} \cdot 0.25 \cdot X_c &= (0.0002 + 0.0005 + 0.0014) \cdot X_c \\ &= 0.0021 \cdot X_c \end{aligned}$$

where N_{li} and N_{ch} are equal to 0. This means that if the published parameters are used, for every kg of COD that disintegrates, 0.1 mole of N is created (5% more than was originally there). As an elemental balance is critical, the benchmark implementation has adopted new values for $f_{XI,xc}$ (= 0.2) and $f_{li,xc}$ (= 0.3). Also, specifically for the BSM2 implementation N_{SI} and N_{XI} have been modified to $0.06/14 \approx 0.00429$ kmol N·(kg COD)$^{-1}$ to be consistent with the ASM1 model default parameter values. Because of this modification, N_{xc} is set to $0.0376/14 \approx 0.00269$ kmol N·(kg COD)$^{-1}$ to maintain the nitrogen balance.

ADM1 contains state variables for inorganic carbon and inorganic nitrogen. These could be used to act as source or sink terms to close mass balances. However, the published stoichiometric matrix does not take this into account. For example, decay of biomass (processes 13–19) produces an equal amount of composite material on a COD basis. However, the carbon contents of these two states are not necessarily identical. A similar problem exists for nitrogen. Batstone *et al.* (2002) suggested that the nitrogen content of bacteria (N_{bac}) should be 0.00625 kmol N·(kg COD)$^{-1}$, which is three times higher than the suggested value for N_{xc}. The fate of the excess nitrogen must be accounted for. The same principle holds for carbon during biomass decay. A similar inorganic carbon modification to the stoichiometric matrix is needed for the processes 'uptake of LCFA (long chain fatty acids), valerate and butyrate' as well as for the disintegration and hydrolysis processes (both carbon and nitrogen).

The recommendation is to include stoichiometric relationships for all 19 processes regarding inorganic carbon and inorganic nitrogen. In many cases the expressions will be zero (depending on the selected values of the stoichiometric parameters) but these modifications will guarantee that the mass balances for COD, carbon and nitrogen are assured. These modifications represent a necessary change to the original ADM1.

The implications of the above changes are not necessarily obvious. The original ADM1 implementation has a carbon content for X_c of 0.03 kmol C·(kg COD)$^{-1}$. However, if we examine the carbon contents of the products arising from disintegration of one COD mass unit of composite material (based on the new fractionation parameters defined above, that is, $f_{XI,xc} = 0.2$ and $f_{li,xc} = 0.3$) and carbon content based on Batstone *et al.* (2002) we get:

$$0.03 \cdot 0.1 \cdot X_c + 0.03 \cdot 0.2 \cdot X_c + 0.0313 \cdot 0.2 \cdot X_c + 0.03 \cdot 0.2 \cdot X_c + 0.022 \cdot 0.3 \cdot X_c = 0.02786 \text{ kmol C} \cdot (\text{kg COD})^{-1} \text{ from } X_c$$

It follows that a significant amount of carbon 'disappears' as a result of disintegration (7%). If the model is updated by adding the stoichiometric relationships to guarantee mass balances of carbon and nitrogen (described above), this disappearing carbon will end up as inorganic carbon and eventually it will lead to carbon dioxide in the gas phase. If the model is not updated as discussed above then 7% of the carbon content of composite material will simply be removed and the carbon mass balance will not hold. Moreover, as this extra carbon eventually ends up as carbon dioxide in the gas phase, the unbalanced original ADM1 will incorrectly calculate the carbon dioxide and methane fractions in the off-gas. To avoid the above problem, a value of 0.02786 kmol $C \cdot (\text{kg COD})^{-1}$ has been used for BSM2 to describe the carbon content of composite material.

4.2.2.2 Acid-base equations

The acid-base equilibrium equations play an important role in ADM1 (e.g., for pH calculations), but Batstone *et al.* (2002) focused more on how the implementation should be done by implicit algebraic equations and were not completely clear on the implementation using ordinary differential equations (ODEs). The general model matrix describes the transformations of valerate ($S_{va,total}$), butyrate, propionate, acetate, inorganic carbon and inorganic nitrogen. However, all these substances are made up by acid-base pairs (e.g., $S_{va,total} = S_{va-} + S_{hva}$). Batstone *et al.* (2002) suggested that when using ODEs, equations for each acid and base component should be defined. However, the BSM2 experience suggests that a better approach is to implement the ODEs based on the total and one of the acid-base components instead. The choice of an ODE or a differential algebraic equation (DAE) system for modelling the pH should not affect the overall results of the model, but the key is to use rate coefficients $k_{A,Bi}$ (where index i indicates any acid-base, i.e., valerate, butyrate, propionate, acetate, inorganic carbon and inorganic nitrogen) for the ODE system that produce the same results as the DAE. Batstone *et al.* (2002) suggested using coefficients that are at least one order of magnitude faster (larger) than the fastest time constant in the remaining system but the benchmark work has shown that the coefficients need to be larger than that and values of $1 \cdot 10^{10}$ $M^{-1} \cdot d^{-1}$ are more appropriate.

4.2.2.3 pH inhibition equations

Batstone *et al.* (2002) use switching functions to account for inhibition due to pH. These functions are, however, discontinuous and can lead to numerical instabilities. To reduce this risk, the BSM2 implementation uses a continuous Hill inhibition function based on the hydrogen ion concentration as first suggested by Siegrist *et al.* (2002). This solution gives the following expressions:

$$I_{pH,aa} = \frac{K_{pH}^{n_{aa}}}{S_{H^+}^{n_{aa}} + K_{pH}^{n_{aa}}} \quad \text{with } K_{pH} = 10^{-\frac{pH_{LL,aa}+pH_{UL,aa}}{2}} \quad \text{and} \quad n_{aa} = \frac{3.0}{pH_{UL,aa} - pH_{LL,aa}}$$

$$I_{pH,ac} = \frac{K_{pH}^{n_{ac}}}{S_{H^+}^{n_{ac}} + K_{pH}^{n_{ac}}} \quad \text{with } K_{pH} = 10^{-\frac{pH_{LL,ac}+pH_{UL,ac}}{2}} \quad \text{and} \quad n_{ac} = \frac{3.0}{pH_{UL,ac} - pH_{LL,ac}}$$

$$I_{\mathrm{pH,h2}} = \frac{K_{\mathrm{pH}}^{n_{\mathrm{h2}}}}{S_{\mathrm{H^+}}^{n_{\mathrm{h2}}} + K_{\mathrm{pH}}^{n_{\mathrm{h2}}}} \quad \text{with } K_{\mathrm{pH}} = 10^{-\frac{\mathrm{pH_{LL,h2}}+\mathrm{pH_{UL,h2}}}{2}} \quad \text{and} \quad n_{\mathrm{h2}} = \frac{3.0}{\mathrm{pH_{UL,h2}} - \mathrm{pH_{LL,h2}}}$$

where subscript LL stands for lower limit and UL for upper limit. To fit the original function given in Batstone *et al.* (2002), $n_{\mathrm{aa}} = 2$ for $I_{\mathrm{pH,aa}}$ and $n_{\mathrm{ac}} = 45$ and $n_{\mathrm{h2}} = 3$ for $I_{\mathrm{pH,ac}}$ and $I_{\mathrm{pH,h2}}$ respectively were chosen. As the appropriate values of n are dependent on the values of $\mathrm{pH_{LL}}$ and $\mathrm{pH_{UL}}$, it is wise to implement n as suggested above. If the value of $\mathrm{pH_{LL}}$ or $\mathrm{pH_{UL}}$ is changed then the corresponding value of n will automatically be adjusted.

4.2.2.4 Gas phase equations

The gas flow rate and partial pressures are essential output variables from the model. However, based on the default procedure in the original model, a potential problem arises when calculating the gas flow and this may lead to numerical problems. Multiple steady states as well as numerical instabilities have been reported among some users. An alternative way of calculating the gas flow is also given in Batstone *et al.* (2002):

$$Q_{\mathrm{gas}} = k_{\mathrm{p}}(P_{\mathrm{gas}} - P_{\mathrm{atm}})$$

where $P_{\mathrm{gas}} = p_{\mathrm{gas,h2}} + p_{\mathrm{gas,ch4}} + p_{\mathrm{gas,co2}} + p_{\mathrm{gas,h2o}}$

This expression assumes an overpressure in the headspace, which is fine, but to compensate for this, the expression needs to be rewritten to obtain the gas flow rate at atmospheric pressure:

$$Q_{\mathrm{gas}} = k_{\mathrm{p}}(P_{\mathrm{gas}} - P_{\mathrm{atm}}) \cdot \frac{P_{\mathrm{gas}}}{P_{\mathrm{atm}}}$$

Although this compensation factor is included, this expression and the original expression do not yield identical results. Depending on the operational overpressure, which is a function of the value of parameter k_{p}, the alternative expression results in a slightly lower gas flow. The reason for this is that the liquid-gas transfer rates ($\rho_{\mathrm{T,8}}$, $\rho_{\mathrm{T,9}}$, $\rho_{\mathrm{T,10}}$) are not identical in the two options. Nevertheless, a comparison of the two expressions when the same overpressure is applied shows very similar results (the relative difference obtained is in the range of $1 \cdot 10^{-5}$). For BSM2, the alternative method (which assumes an overpressure in the headspace) is used. The reader is cautioned to note that if the physical or operational conditions of the digester model are changed (volume, load, …), for example if applying the ADM1 as a stand-alone model outside the framework of BSM2, then the parameter k_{p} will have to be adjusted to achieve a reasonable overpressure in the headspace.

4.2.2.5 DAE simplifications and simulation speed

BSM2, by definition, has a number of requirements including the ability to handle dynamic inputs, time-discrete and event-driven control actions as well as stochastic inputs or noise. All of these simulated events must be efficiently handled with a reasonable simulation speed. Initial ADM1 implementations showed that the ODE implementation is problematic for use in the BSM2 framework due to the computational effort and the stiffness of the model. A system is referred to as being stiff when the range of the model time constants is large. This means that some of the system states react slowly and others react very quickly. ADM1 is a *very* stiff system with time constants ranging from fractions of a second to months.

This makes the simulation challenging and in order to avoid excessively long simulation times, creative implementations of the model are required.

Some numerical solvers are stiff solvers and, consequently, capable of solving stiff systems efficiently. However, stiff solvers struggle to handle dynamic input including noise. The more stochastic or random an input variable is, the more problematic the numerical simulation becomes for a stiff solver. BSM2, which includes ASM1 and ADM1 models, is a very stiff system and, consequently, a stiff solver should be used. However, as BSM2 is a control simulation benchmark, noise must be included, calling for an explicit (i.e., non-stiff) solver. By rewriting an ODE system into a DAE system, the stiffness can be decreased, allowing for explicit solvers to be used and for stochastic elements to be incorporated.

In Batstone *et al.* (2002), it is suggested that the pH (S_{H+}) state be calculated by algebraic equations, assuming instantaneous equilibrium. However, this will only partially solve the stiffness problem. There are other fast state variables in the model and a closer investigation shows that the state describing hydrogen (S_{h2}) also needs to be approximated by an algebraic equilibrium equation in order to enhance the performance when simulating ADM1 using an explicit solver. As a result, two different DAE models (Rosen *et al.* 2006; Technical Report No. 5) have been developed: a model based on algebraic pH (S_{H+}) calculations (DAE_{pH}) and a model based on algebraic pH and S_{h2} calculations ($DAE_{pH,S_{h2}}$).

An implicit algebraic equation for the pH calculation is given in Batstone *et al.* (2002). In the BSM2 case, the Newton-Raphson method used in Volcke *et al.* (2005) was implemented for the calculation of the pH and equilibrium concentrations (i.e., the acid-base equations). The reader is referred to Volcke *et al.* (2005) for details on the method, but in essence the iterative procedure is repeated until a predefined tolerance value (10^{-12} for BSM2) is met. Normally only two or three iterations are required to solve the equation at each integration step. The differential equation for the S_{h2} state can be approximated by an algebraic equation using the same principle of simply setting its differential to zero (assuming infinitely fast dynamics) and the iteration is carried out in a similar way.

In order to verify the DAE implementation suggested here, several numerical tests were undertaken. At steady state, the tests revealed very small (close to machine numerical precision) differences. Only minor errors were encountered in the pH-value with the largest relative errors in the range of 10^{-6}. The differences during dynamic conditions have been studied extensively as part of the benchmarking work and have been found to be fully acceptable. Some small differences are unavoidable due to the numerics of the ODE and DAE implementations, but these differences are well within a reasonable tolerance. The main difference was in the hydrogen state where the mean relative error was slightly larger than 0.01%. Although not a true state in the model, the gas flow rate seems to be highly sensitive to the integration algorithm and the time step used. The use of some solvers resulted in 'nervous' behaviour in the gas flow rate with noise in the range of a few percent. This sensitivity appears in all model implementations when noise is present, and is most likely caused by integration numerics.

The simulation speed was tested using different solvers. The results are shown in Table 4.5 as relative simulation times. It is interesting to note that the implementation of a DAE pH solver alone does not give any significant improvement in simulation speed. Substantial improvement is not obtained unless S_{h2} is algebraically solved also. An investigation into the relative contributions to the simulation speed, revealed that both pH and S_{h2} must be removed to significantly reduce the stiffness of the BSM2 defined system.

The choice of implementation is up to the user. If acceptable computation times can be achieved with the ODE implementation, there may be no advantage to using a DAE option. However, for Matlab/Simulink and WEST, it appears that the $DAE_{pH,S_{h2}}$ option is the only practically feasible choice for BSM2.

Table 4.5 Relative simulation times for the three different implementations of ADM1 in Matlab with different solvers (Rosen *et al.* 2006).

	ODE45	ODE23	ODE23tb (stiff)	ODE15s (stiff)
ODE	53	96	28	18
DAE_{pH}	53	85	27	17
$DAE_{pH,S_{h2}}$	1 (ref.)	1.2	28	18

4.2.2.6 Model parameters

The mesophilic ADM1 parameters used for BSM2 are listed in Appendix A. In cases where the parameter value in BSM2 differs from the default ADM1 parameter value (Batstone *et al.* 2002) the default value is also given. For BSM2 purposes, the values produce satisfactory results, but the Task Group is not suggesting these parameter values be used for any other purpose. It should be noted, for instance, that the parameter values for the hydrolysis rates (10 d^{-1}) used in both ADM1-BSM2 and published as default parameter values in the ADM1 STR are nowadays considered to be at least ten times too large (Batstone 2002–2008, personal communication).

4.2.3 ASM/ADM interfacing

The development of BSM2 is in keeping with the recent trend in WWTP modelling towards plant-wide or whole plant models and does not merely focus on one specific part of the process. The inclusion of these different subunits can cause conceptual and mathematical difficulties within the model especially when the individual subunit models were developed in isolation and the models do not use a common set of state variables. In the cases when the subunit models use different sets of state variables, conversion algorithms (or interfaces) are required to convert model variables from one set to the other and *vice versa* when necessary. These algorithms are widely used and typically quite straightforward, but the process becomes complicated when different biological models are interfaced, which is the case when an activated sludge unit is coupled to an anaerobic digestion process as is the situation in BSM2.

One of the key challenges in the BSM2 development was the definition of an interface between the activated sludge portion of the model (ASM1) and the anaerobic digestion portion (ADM1). The development of the interface described here required several iterations, but the end result is an interface that combines quantitative understanding with conceptual integrity of the state variable meanings. The interface balances COD, nitrogen and charge while creating ADM1 inputs that are consistent with the biodegradability of anaerobic digester feeds observed in practice. The interface ultimately agreed upon is described in limited detail here and although this interface has been designed for use with BSM2, it is widely applicable to ADM1 input characterisation in general.

As part of the BSM2 evolution, Copp *et al.* (2003) proposed an interface based solely on the conservation of both COD and TKN. The key feature of this ASM → ADM algorithm was the maximisation of X_c, the state variable representing complex material in ADM1. However, an analysis of the anaerobic results obtained using this interface revealed several key shortcomings: 1) the difference between the methane produced per unit VSS applied for primary and secondary sludges was too small as compared to the literature (Gosset & Belser, 1982; Siegrist *et al.* 2002; Svärd, 2003); and, 2) the digestion model produced an inert fraction that was unrealistically high regardless of the input. As a result a new, more comprehensive approach was adopted for BSM2 (Nopens *et al.* 2009).

4.2.3.1 ASM1 to ADM1 conversion

The steps in the BSM2 interface to convert state variables from ASM1 to ADM1 are depicted in Figure 4.7 and the interface parameters for the various steps are listed in Appendix A. Step 1 involves the removal of oxygen and nitrate through the corresponding elimination of an equivalent mass of degradable COD using first soluble substrate, then particulate substrate if necessary. The purpose of step 2 is to maximise the amino acid input using the available S_{ND} and S_S from ASM1 as constraints. In step 3, the protein input is maximised using X_S and X_{ND} as constraints with lipids and carbohydrates being used to mass balance any left over COD. For the BSM2 implementation the left over COD is split 70/30 (X_{li}/X_{ch}) for primary sludges and 40/60 for secondary sludges (based on Siegrist *et al.* 2002). If there is insufficient X_S to convert all of the X_{ND} to protein, the excess X_{ND} is allocated to the generalised TKN pool.

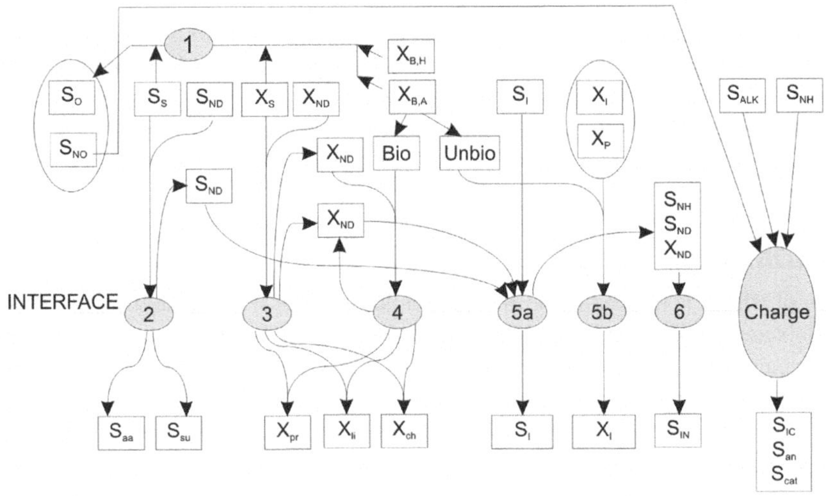

Figure 4.7 Pictorial representation of the ASM to ADM interface. Computational steps are shown as numbers, with inputs and outputs from each step arrowed (from Nopens *et al.* 2009).

In step 4, the incoming biomass is converted into proteins, lipids, carbohydrates and inerts. It is assumed during this step that a portion of the biomass (32%) is non-biodegradable and is removed from the incoming pool of biodegradable COD. The nitrogen associated with the incoming biomass and any remaining X_{ND} from step 3 is assumed to be protein-associated nitrogen. This mass of COD is removed from the remaining pool of biomass COD and any COD still remaining is mapped to lipids and carbohydrates. Step 5 involves the mapping of the ASM1 particulate inerts (X_I, X_P and the non-biodegradable biomass COD-fraction) to the ADM1 particulate inert state variable. It is assumed for this process that aerobically inert material is also anaerobically inert (Ekama *et al.* 2007). The mapping is easily done provided the nitrogen content of the inerts is the same in both ASM1 and ADM1 as is the case in BSM2. When the nitrogen contents are not the same, special rules have to be applied to ensure mass balances for both COD and TKN. Step 6 maps any remaining nitrogen (S_{ND}, X_{ND} and S_{NH}) to the inorganic nitrogen state variable and the conservation of COD and TKN is complete.

Calculation of the physicochemical system, including inorganic carbon (S_{IC}), and the cation/anion concentration is the final step. This is done in a two-step process, by first balancing weak acid and base charges across the interface, and then calculating cation or anion concentrations by setting the interface pH. S_{IC} is calculated by balancing weak acid and base charges across the interface. All of the charge equation components are known or calculable except the inorganic carbon, which provides the necessary degree of freedom to ensure charge continuity across the interface. Finally, as the interface balances charge via S_{an} and S_{cat}, a full charge balance can be calculated and the appropriate values assigned.

Calculating a balanced ionic input for ADM1 can be a challenge, given variations in bicarbonate, ammonia, organic acids and other buffers. The interface model proposed here removes much of this difficulty. This approach to charge balancing removes much of the guess work in equalising charges across the interface and enables the states such as S_{IC}, S_{an} and S_{cat} to be consistently set given a target input pH. BSM2 assumes a neutral ASM1-side pH (as it is not calculated) but the interface can be easily adapted to allow for variable pH.

4.2.3.2 ADM1 to ASM1 conversion

The different steps in the ADM1 to ASM1 interface are depicted in Figure 4.8. In the first step, biomass is converted to X_P and X_S taking into account the available biomass nitrogen and the corresponding nitrogen requirements of the products. If insufficient nitrogen is available inorganic nitrogen is used to satisfy the nitrogen requirement. Next, the remainder of the particulate COD (X_c, X_{pr}, X_{li}, X_{ch}) is mapped directly to X_S with the nitrogen associated with the proteins (X_{pr}) being mapped to X_{ND}. In the third step, the particulate inerts are directly mapped to X_I. Step 4 involves mapping soluble inerts with the associated nitrogen being portioned to ammonia. In step 5, all the soluble COD-states (S_{su}, S_{aa}, S_{fa}, S_{va}, S_{bu}, S_{pro} and S_{ac}) are mapped to soluble substrate COD (S_S) and the nitrogen associated with the amino acids is mapped to S_{ND}. As a final step, the charge balance is used to calculate the alkalinity.

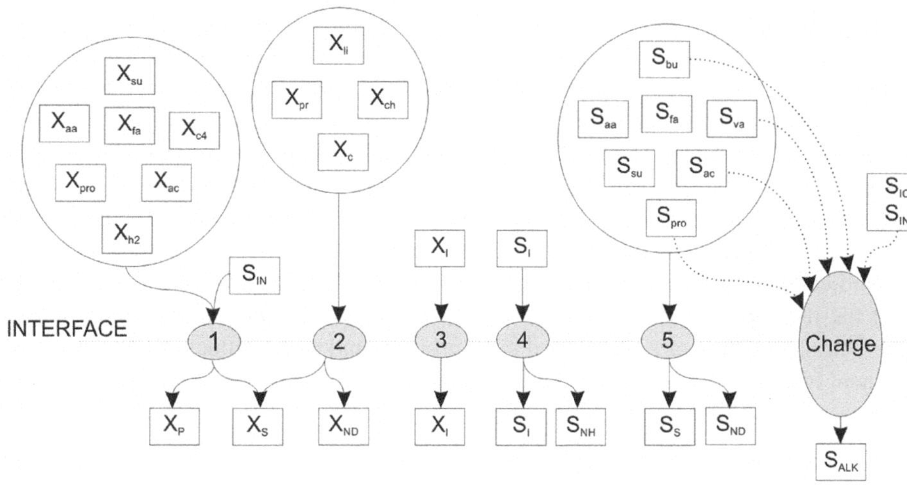

Figure 4.8 Pictorial representation of the ADM to ASM interface. Computational steps are shown as numbers, with inputs and outputs from each step arrowed (from Nopens *et al.* 2009).

4.2.3.3 Further remarks

Interfacing activated sludge with anaerobic digestion and addressing the general problems associated with the composite material in ADM1 have been some of the most intensively addressed topics since

ADM1 was published (e.g., Copp *et al.* 2003; Huete *et al.* 2006; Kleerebezem & van Loosdrecht, 2006; Vanrolleghem *et al.* 2005b; Volcke *et al.* 2006b; Wett *et al.* 2006). There are several important issues that have been addressed with this BSM2 interface. The Task Group has determined that maximising the use of X_c does not produce a distinction between primary and secondary sludges in the influent and thus imposes secondary sludge degradation kinetics on the more degradable primary sludge. Furthermore, when the lumped variable X_c is used, the impact that these upstream operational items impose can be masked and the true degradability of the influent might be misinterpreted. The better approach, as proposed, is to map the state variables directly to the constituent compounds (proteins, carbohydrates, lipids, amino acids, sugars and VFA) and bypass the X_c state variable altogether. Using the interface proposed by the Task Group should provide a more realistic estimate of the sludge degradability.

By mass balancing organic nitrogen, COD, charge, carbon and particulate/soluble fractions where possible, the interface balances practicality and reasonable kinetic behaviour. Use of an interface model, such as the one proposed here can allow others to transparently and consistently translate inputs to ADM1. Although designed for primary and secondary sludges, the methodology is applicable to any feed regardless of its source (e.g., Zaher *et al.* 2009). The source code for the interface is available as part of the STR supplemental material.

4.2.4 Solids separation models

4.2.4.1 Primary clarifier

The primary clarifier model proposed by Otterpohl (1995) – hereafter referred to as the Otterpohl model – provided a reasonable representation of the primary settling process for BSM2 (Technical Report No. 4). It was not the purpose to choose a model that described the primary settling process in detail. Rather, the main driver for the choice of this model was the ability of the model to produce a reasonable estimate for the streams over the weir (activated sludge influent) and in the underflow (anaerobic digestion influent). Although empirical in nature, it was determined that this model was able to produce reasonable streams with respect to the concentrations of particulate material in both streams.

The Otterpohl model describes a single homogeneous tank with an empirical separation (Figure 4.9) of the output into primary effluent (overflow) and primary sludge (underflow) streams. The original model description does not consider the primary sludge stream, but the necessary mass balance calculations were added to the BSM implementation. The homogeneous tank concept accounts for the hydraulic retention time and concentration smoothing in the clarifier and the empirical separation equation, which is a function of the hydraulic retention time, is used to calculate the state variable concentrations in the effluent stream. The sludge flow rate is assumed to be proportional to the influent flow and a mass balance on each component is used to calculate the state variable concentrations in that sludge flow.

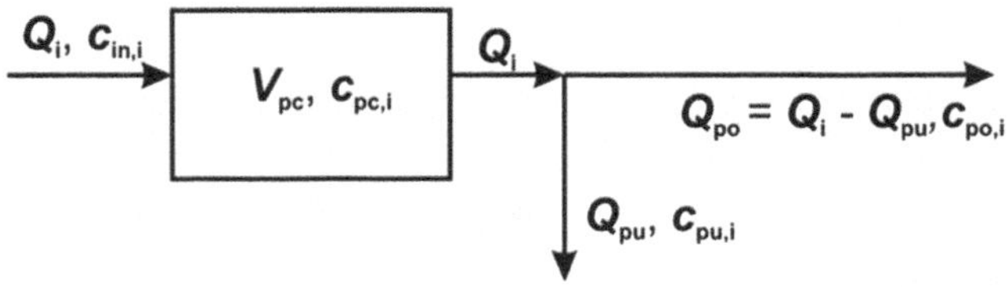

Figure 4.9 General overview of the Otterpohl (1995) model assumptions (pu – primary underflow; po – primary overflow).

In the model, the COD removal efficiency η_{COD_p} [%] (i.e., the empirical solids separation) can be calculated for every time step using:

$$\eta_{COD_p}(t) = f_{corr}(2.88 f_X - 0.118) - (1.45 + 6.15 \ln(t_h(t) \cdot 24 \cdot 60))$$

where f_X is the mean fraction of particulate COD [–], $t_h(t)$ is the hydraulic retention time [d] and f_{corr} is a dimensionless correction factor for the COD removal efficiency in the primary clarifier. It should be noted that $t_h(t)$ is not a constant but a dynamic variable which changes when the influent flow varies. However, the parameter f_X is assumed to be constant. The instantaneous hydraulic retention time, $t_h(t)$ is the result of a first-order smoothing filter with a time constant of 3 hours with Q_{in} as an input. The primary effluent concentration of each state variable is calculated using:

$$f_i = 1 - \frac{\eta_{COD_p}}{100} f_{SX,i}$$

where $f_{SX,i}$ is equal to 0 for all soluble components and something greater than 0 for all particulate fractions. This formulation gives the user the ability to control the settling of each particulate state variable individually. In the benchmark case, all the particulate state factors were set to 1. One idea that has been put forth is that $f_{SX,XS}$ (the settling factor for X_S) could be set to 0.5 to account for the portion of X_S that does not settle. Although defined as a particulate component, it is widely accepted that X_S includes a colloidal component that does not readily settle in the primary clarification process (Henze *et al.* 1987). So far, this has not been implemented in BSM2.

4.2.4.2 Secondary clarifier

As with the biological process models, international acceptability was the overriding criterion for choosing a secondary settling model. In this case a biologically inactive process is modelled using the double-exponential settling velocity function of Takács *et al.* (1991). Even though recent publications have presented alternatives (Bürger *et al.* 2011; Plósz *et al.* 2011), the Takács model is still internationally accepted, in part, because it is applicable to both hindered and flocculent settling conditions. The model is implemented in the predominant simulation software packages and widely used in practice making it the best choice for the BSMs.

The benchmark implementation assumes a 10-layer system (Figure 4.10) with the settling behaviour and solids in any one layer calculated as the net effect of settling plus the solids movement due to liquid flow. The layers are numbered from the bottom (layer 1) with the sixth layer being designated as the feeding layer. In each layer the mass balance for the sludge takes into account the solids flux due to settling and the transport of solids into and out of the layer due to the flow of liquid. Above the feed layer (layers 7–9), the upward flow carries solids upwards (to the layer above and from the layer below) and settling carries solids downward (from above and to below). Below the feed layer, both the liquid flow and particle settling contribute to the downward movement of solids. The feed layer is a special case with solids entering the layer from the feed, and exiting the layer both in an upward and downward direction due to the flows. Solids are removed from the clarifier both from layer 1 (RAS) and layer 10 (effluent) ensuring a solids mass balance across the clarifier.

Settling due to gravity in each layer is calculated through the solids flux, which is defined as the product of the settling velocity, $v_s(X_{sc})$, and the solids concentration in that layer, (X_{sc}), as:

$$J_s = v_s(X_{sc}) X_{sc}$$

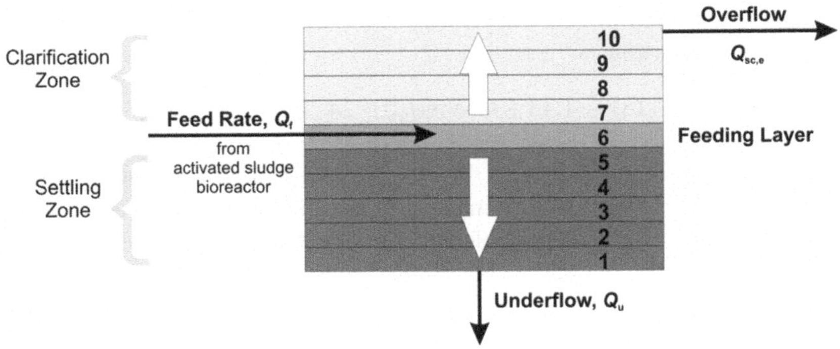

Figure 4.10 Schematic model of the secondary clarifier.

The double-exponential settling velocity function that is used for determining the particle settling velocities inside each layer uses a number of parameters and these have been fully defined in Appendix A. The general equation is:

$$v_{sj} = v_o e^{-r_h X_j^*} - v_o e^{-r_p X_j^*}$$

$$0 \leq v_{sj} \leq v_o'$$

where v_{sj} is the settling velocity in layer j [m · day^{-1}], X_j^* is the solids concentration in layer j [g · m^{-3}], subject to the limiting condition $X_j^* = X_j - X_{min}$, X_j is the suspended solids concentration in layer j [g · m^{-3}], and X_{min} is the minimum attainable suspended solids concentration [g · m^{-3}] calculated as $X_{min} = f_{ns} \cdot X_{in}$ (X_{in} is the mixed liquor suspended solids concentration entering the settling tank and f_{ns} is the non-settleable fraction). Figure 4.11 shows the settling velocity curve that is obtained with the benchmark parameter values.

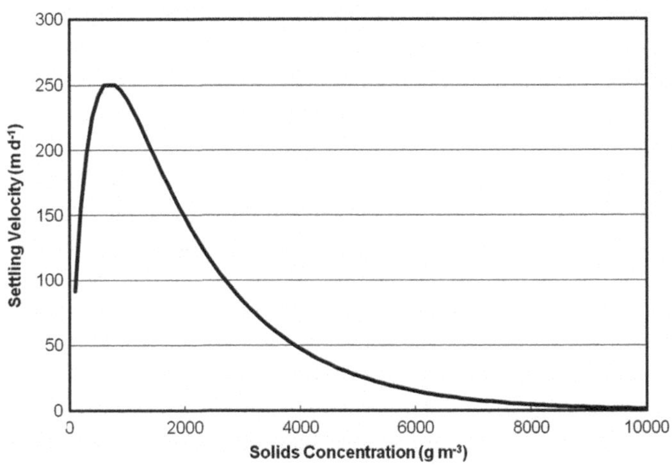

Figure 4.11 Settling velocity curve obtained using the benchmark parameter values.

As the benchmark implementation assumes a non-reactive model, the mass balances are applied to the calculated incoming total suspended solids (TSS) and not to the individual particulate ASM state variables.

Because of this, the output from the model is a TSS value for each layer. However, to transform the output into the output streams, the calculated TSS values must be converted into the individual particulate ASM state variables. To do this, it is assumed that the incoming fractionation (i.e., the fractionation of the feed flow to layer 6) is directly propagated to the fractionation of the secondary clarifier output streams. That is, the particulate fractions in the effluent and underflow are assumed to be the same as the influent fractions. The Task Group recognises that this is a simplification, but consistent with current clarifier modelling practice.

NOTE: Even though the benchmark models are fully defined and all the model parameters are specified, users should always be aware of simulator-specific issues that can have an impact on the model output. An example of this is shown in Figure 4.12 where the initial dynamic output of three different simulators is shown. The differences illustrated in the figure are the result of the different means used to propagate soluble components through the secondary clarifier. In these three instances the particulate components are modelled in precisely the same way, but the soluble components are modelled differently. The output differences are a direct result of the variation in the number of settler layers used for soluble components. At steady state, these differences make no difference in the model output, but the effect becomes apparent under dynamic conditions. The official benchmark definition states that soluble state variables should be modelled using 10 layers. The issue was identified during ring-testing and the commercial simulators subsequently prepared solutions to address this issue, but users are cautioned to check their implementation. The curves presented in Figure 4.12 were produced by three different simulators, but it should be made clear that each simulator produced the same result when the number of soluble layers was identical.

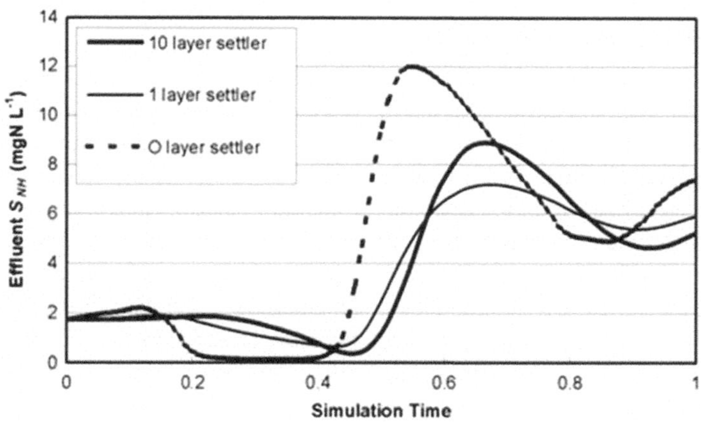

Figure 4.12 Soluble component produced by the same settler model but with a different number of soluble layers. The 10-layer soluble model is defined in the benchmark. Note that all the simulators produced the same result when the number of soluble layers was identical (from Copp, 2002).

4.2.4.3 Thickener

A thickener is included in BSM2 but it was not deemed to be a critical unit process requiring a comprehensive model. Rather, the thickener in BSM2 is modelled as an ideal unit with no volume that 'thickens' the wasted sludge from the secondary clarifier to reduce the volume going to the digester. The unit process model assumes a fixed sludge concentration in the underflow (7%) and a fixed removal efficiency (98%). With these two constraints, the model calculates the sludge flow rate. Mass and flow balances are then used to calculate the concentrations and liquid flow over the weir. The concentrations of the soluble components do not change through the thickener model.

4.2.4.4 Dewatering unit

The dewatering unit in BSM2 is used to reduce the volume of the digested sludge prior to disposal. No assumption is made about the type of equipment, but as with the thickener, this unit is assumed to be ideal with no volume. In this case the output sludge concentration is assumed to be constant at 28% and the removal efficiency is again assumed to be 98%. As with the thickener, mass and flow balances are used to calculate the output flows and concentrations using those two constraints.

4.2.5 Reject water storage tank

The reject water from the dewatering unit can be recycled to a non-reactive storage tank prior to being recycled to the process stream. The storage tank behaviour depends upon the flow rate from the dewatering unit, the available storage volume and the fate of the stored reject water. It is assumed that the effluent flow from the storage tank will be equal to the influent flow if the tank is full, equal to a defined pumped flow if less than full and equal to the influent flow if completely empty. The effluent flow should not exceed a reasonable rate, to avoid unrealistic instantaneous emptying of the tank and the flow is directed to either the primary clarifier influent or effluent as defined by the user.

4.3 SENSORS AND ACTUATORS

Simulation provides a unique opportunity to test control options in a virtual world, but any detailed investigation should consider as many real issues as possible. Sensors and actuators are a case in point. It might be a valid option to use ideal sensors (no noise, no delay) to evaluate the theoretical potential of a control strategy. It might also be reasonable to use ideal measurements for flows and pump controls. However, to avoid unrealistic control behaviour, the dynamic behaviour of sensors and actuators (control handles) as well as additional measurement noise must be considered. Delays introduced by the propagation of sensor signals and actuators can significantly affect the performance of the control actions, so these delays should be considered during a detailed investigation. The mathematical description of these things for the BSMs has focused on simplicity rather than the complete and accurate reproduction of the true behaviour. This simplified approach adds the necessary complexity without adding an unnecessary computational burden.

Table 4.6 Definition of the variables used in the sensor models.

Variable	Definition
$u(t)$	ideal measurement signal from process
$u_2(t)$	delayed measurement signal (intermediate variable)
$y(t)$	real measurement signal from sensor (delayed, noisy, limited)
τ	first-order time constant
t_r	response time of sensors, equal to 90% time of step response
t_d	delay time of sensor step response
$R_{td/tr}$	ratio of delay time to response time of a sensor
n	system order

4.3.1 Sensors

4.3.1.1 Concept

The sensors defined in the benchmark system are limited to sensor classes. The aim of the sensor classification is to describe different sensor types but also to limit the number of sensor options available. A user configurable class has not been defined as this would make it difficult to compare different benchmark studies. Rather, sensors are classified using 'response' and 'delay' time parameters (Table 4.6). The response time characterises the sensor dynamics based on a step response as presented in Figure 4.13. The response time is the sum of the delay and the rise (or fall) time. The delay is defined as the time to reach 10% of the final value of a step response (t_d). The overall time to reach (and not to leave) a band from 90%–110% of the final value is defined as the response time (t_r). It is assumed that these parameters are known for a given sensor and are sufficient to describe the dynamics of that sensor. Six sensor classes are defined for the benchmark system and the response times characterising each class are listed in Table 4.7. A list of typical sensors and their placement into these categories is also provided in the table. Note that flow measurements in the BSMs are considered to be ideal.

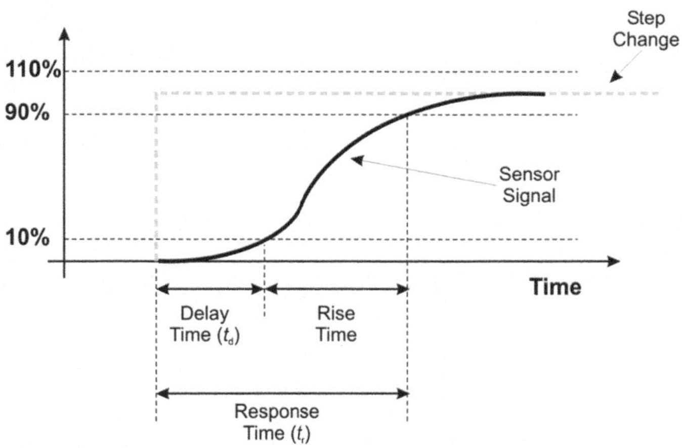

Figure 4.13 Definition of response time.

Table 4.7 Sensor classes according to Rieger *et al.* (2003).

Sensor classes	Response time (t_r) [min]	Measuring interval (t_i) [min]	Examples
Class A	1	0	Ion sensitive, optical without filtration
Class B0	10	0	Gas sensitive + fast filtration
Class B1	10	5	Photometric + fast filtration
Class C0	20	0	Gas-sensitive + slow filtration
Class C1	20	5	Photometric + slow filtration or sedimentation
Class D	30	30	Photometric or titrimetric for total components

Note: A measuring interval equal to 0 means continuous measurement.

Class A is essentially an ideal sensor. The response time of one minute prevents unrealistic control applications. Class B is meant to encompass classic on-line analysers with rapid filtration and short sample loops. Class C sensors describe analysers with slow filtration or sedimentation. Class D includes all batch measurements like respirometry and sensors for total components. To take into account continuous and discrete signals, the classes B and C are divided into two subclasses. Five minutes is selected as the measuring interval for one of these subclasses, which is a typical minimum value for photometric analysers.

In addition to choosing the sensor class, the user must also define the measuring range for each sensor with an understanding that the standard deviation on the sensor signal is assumed to be 2.5% of the maximum measurement value. This standard deviation is used to reflect the measurement noise which is added to the delayed sensor signal by the model. Noise is not modelled exactly, but rather this approach is used to take into account some of its effects. In order to get comparable benchmark simulation results, the sensor noise signal is fully defined. Defining the noise eliminates the choice of a random signal which would complicate the comparison of control strategies due to its random, non-repeatable, nature. The noise signal has a standard deviation of 1, which is then multiplied by the defined noise level (2.5% of the maximum measurement value). The noise is generated as white zero-mean normally distributed noise with a sample time of one minute and linearly interpolated between the samples. Several different default noise signals have been developed (downloadable from www.benchmarkwwtp.org) for use with the BSMs so that different noise signals can be used with different sensors to avoid correlated noise on different sensor signals.

As an example, the oxygen and nitrate sensors used in the default closed-loop test case for the BSMs are described as:

– *oxygen sensor*: Class A, measurement range: 0–10 g $O_2 \cdot m^{-3}$, measurement noise $\delta = 0.25$ g $O_2 \cdot m^{-3}$;
– *nitrate sensor*: Class B_0 with a measurement range 0–20 g $N \cdot m^{-3}$, measurement noise $\delta = 0.5$ g $N \cdot m^{-3}$.

4.3.1.2 Time response

To clarify the implementation of the sensor models, some additional details for continuous sensor signals are outlined here, but the reader is referred to the detailed technical report for a full description.

Let the original sensor signal be $u(t)$. This signal has to be transformed into a delayed signal ($u_2(t)$) with the desired response time (t_r). To create a sensor model with the desired response time and a typical ratio for delay to response time ($R_{td/tr} = t_d/t_r$), a series of first-order delay transfer functions is used. The number of first-order transfer functions in series (n) determines the ratio of delay time and response time (as defined in Figure 4.13).

For the sensor class A, a response time (t_r) of 1 min and a system order of $n = 2$ are suggested. The assumed sensor model, written as a transfer function, is:

$$G_S(s) = \frac{u_2(s)}{u(s)} = \frac{1}{1+\tau s}\frac{1}{1+\tau s}$$

Implementation then involves choosing a τ that gives $t_r = 1$ min. Analysing the step response of this transfer function gives the following relationship:

$$\frac{t_r}{\tau} = 3.89$$

Substitution gives:

$$\tau = \frac{t_r}{3.89} = 0.257$$

Similar to the response time, the delay time is determined by n and τ, and the ratio between delay time and response time ($R_{td/tr} = 0.133$) is only a function of n (=2).

The time response model can be alternatively described by differential equations instead of the Laplace transfer function used above. The dynamic behaviour of the sensor is described below:

$$\frac{dx_1(t)}{dt} = \frac{1}{\tau}u(t) - \frac{1}{\tau}x_1(t)$$
$$\frac{du_2(t)}{dt} = \frac{1}{\tau}x_1(t) - \frac{1}{\tau}u_2(t)$$

For sensor classes with longer response times (i.e., B and C), n becomes a larger number. For class B with a response time of 10 min and for class C with its response time of 20 min, the corresponding transfer function is:

$$G_S(s) = \frac{u_2(s)}{u(s)} = \frac{1}{1+\tau s}\frac{1}{1+\tau s}\frac{1}{1+\tau s}\frac{1}{1+\tau s}\frac{1}{1+\tau s}\frac{1}{1+\tau s}\frac{1}{1+\tau s}\frac{1}{1+\tau s}$$

A value for τ (=t_r/11.7724) was calculated numerically by simulating the transfer function for an input step signal and a given value for τ. The step response was analysed for the resulting response time t_r and the ratio was calculated. This leads to a delay to response time ($R_{td/tr}$) ratio of 0.392. The dynamic behaviour of the sensor using differential equations is described as follows:

$$\frac{d\,x_1(t)}{dt} = \frac{1}{\tau}u(t) - \frac{1}{\tau}x_1(t)$$
$$\frac{d\,x_{k+1}(t)}{dt} = \frac{1}{\tau}x_k(t) - \frac{1}{\tau}x_{k+1}(t); \quad k = 1\ to\ 6$$
$$\frac{d\,u_2(t)}{dt} = \frac{1}{\tau}x_7(t) - \frac{1}{\tau}u_2(t)$$

The step responses for a sensor in classes A, B_0 and C_0 are presented in Figure 4.14.

4.3.2 Actuators

For reasons of simplicity, most available control handles are considered to be ideal with the exception of aeration. In the BSM1 closed-loop test case, only two control handles are used: the internal recirculation flow rate (Q_{int}) and the oxygen transfer rate in reactor number 5 (K_La_5), but in total there are approximately 30 individual control handles that can be manipulated within the defined benchmark plant and this dramatically increases its flexibility (in BSM2 there are approximately 60 control handles). The limitations for the different control handles are defined and listed in Appendix A. A special case exists, however, for the aeration system. The aeration system ($K_La_{1\text{-}5}$) is defined with significant dynamics. A response time of $t_r = 4$ min is defined and the second-order time delay function gives a reasonable model of this process. The time constant for each delay is $\tau = t_r/3.89 = 1.03$ min. Users are encouraged to implement these constraints to ensure more realism into their simulations while at the same time ensuring that any results are comparable with other benchmark results.

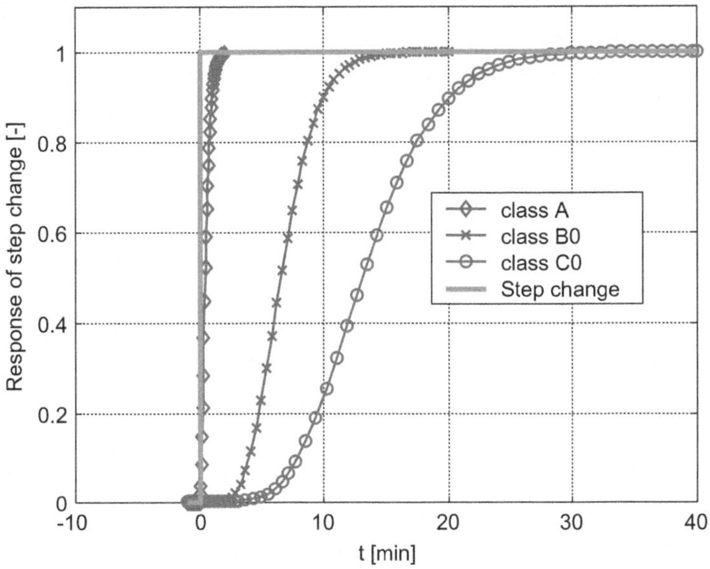

Figure 4.14 Step response of sensor classes A, B_0, C_0.

4.3.3 Faults and failures

The undisputed realisation that sensor and actuator behaviour will have an effect on control performance prompted the investigation of incorporating faults and failures into the BSMs. Sensor and actuator faults and failures are often neglected in simulations for control strategy development and testing, although it is well known that they represent a significant obstacle at full-scale facilities. Faults and failures can affect the performance of a control system and as a result, a significant amount of time is often spent on the 'safety net' around the control system when applied at full scale. Incorporation of faults and failures in simulations can be done simplistically by manually imposing faults at desired locations during the simulation period. However, it is difficult to obtain a realistic disturbance and fault distribution with this approach. So, a framework for incorporating faults and failures based on Markov chains was developed for the benchmark systems (Rosen *et al.* 2008).

A Markov chain contains a number of different states (s_i) that the system switches between according to certain transition probabilities. In its simplest form, a sensor (or actuator) is modelled having two states. One state (s_1) represents a fully functional sensor and the other state (s_2) represents any sensor fault. A Markov chain for this problem is depicted in Figure 4.15. In this case, the transition probability p_{12} defines the probability for a fault at any given time instance given that the sensor is functioning. The probability p_{11} defines the probability that the sensor will remain functioning as is, naturally, $p_{11} = (1 - p_{12})$. Conversely, p_{21} defines the probability that the sensor will be repaired (or replaced) and p_{22} defines the probability that the sensor will remain broken.

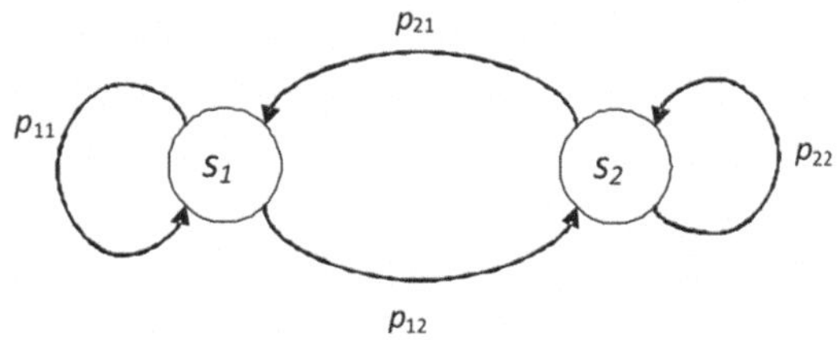

Figure 4.15 A Markov chain with two states.

More generally, a Markov chain is described by its probability matrix, **P**. By defining the desired stationary solution (e.g., for a three-state system *(operational, failure, calibration)* this could mean $\mathbf{p}^* = [0.95\ 0.02\ 0.03]$, that is, working properly 95% of the time, failing for 2% and being calibrated for the remaining 3%) and solving, it is possible to calculate the transition probabilities of **P**. The stationary solution represents the average behaviour of the sensor. The discrete dynamics of the system over time are calculated as $\mathbf{p}(k) = \mathbf{p}(0) \cdot \mathbf{P}^k$ where $\mathbf{p}(0)$ is the starting condition and k is a counter. The Markov approach has an advantage as it is possible to tune the model to behave in a desired and realistic manner.

For the benchmark, it is assumed that a sensor (or actuator) can only be in one state at a time, (i.e., multiple states are neglected, a fundamental prerequisite for all Markov chain systems) and the sensor state types are as follows: (1) *operational (with noise)*; (2) *drift;* (3) *shift (off-set)*; (4) *fixed value*; (5) *complete failure*; (6) *incorrect gain*; and (7) *calibration*. Illustration of these states is shown in Figure 4.16. As can be seen in the figure, the transitions between sensor states vary. States 2–5 can only become active when the sensor is operating (i.e., in state 1). States 4 and 5 return to state 1 with the probability of p_{41} and p_{51}, respectively, whereas states 2 and 3 can only move to state 7 (calibration) with the probability of p_{27} and p_{37}, respectively. State 6 can only be activated as a direct consequence of calibration (p_{76}) and is only ended by a new calibration (p_{67}). From these assumptions, a probability matrix can be defined.

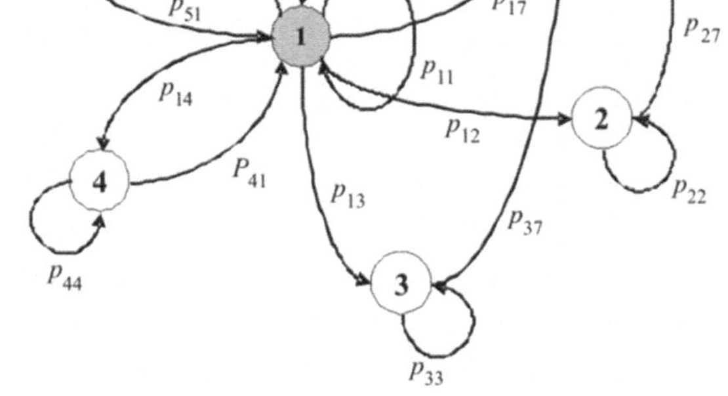

Figure 4.16 A Markov chain representation of the sensor fault model (Rosen *et al.* 2008).

The realisation of a Markov chain model as a function of time is straightforward and can be written in just a few lines of code. For each time step, the transition to a new state is done according to a uniformly distributed random number [0 ... 1], which is compared with the transition probabilities of the particular state the system is currently in.

To illustrate how the model can be tuned to display the behaviour of a specific sensor, online ammonia measurements will be used to illustrate the approach. Real ammonia measurements covering a period of 286 days and sampled every 15 minutes (in total 27503 samples) are shown in Figure 4.17.

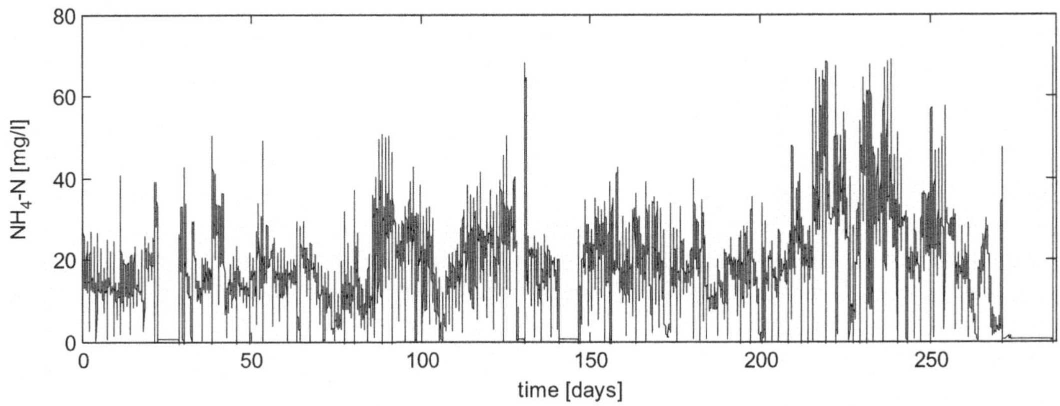

Figure 4.17 Ammonia measurements in the influent of a full-scale WWTP.

The figure clearly shows at least 3 major sensor breakdowns: at days 22–29, 141–146 and 271–286 and a closer inspection reveals another 12 breakdowns. These are classified as state 5 – *complete failure.* The sensor is calibrated (or maintained, e.g., cleaning) 102 times during the period and these events are classified as state 7 – *calibration.* Faulty calibration instances can be seen at days 38–42, 85–92 and 130–131 and are classified as state 6 – *incorrect gain.* States 2, 3 and 4 are too difficult to distinguish based on this figure and are not considered in this example.

The second step towards characterising the measurement data is to determine the average time that the sensor is in any one state. The *average* duration of the 15 complete breakdowns (type 5) is 251 samples (approx. 2.6 days). On average, the sensor has an incorrect gain after calibration (type 6) for 354 samples (approx. 3.7 days) and takes an average of 3.8 samples (approx. 1 hour) for calibration (type 7). The last piece of information needed is the fraction of the total time that each state occurs. The sensor is in the type 5 state for 14% of the time. The sensor is in the type 6 state 4% of the time and in the type 7 state 1% of the time. As states 2, 3 and 4 are not considered, the sensor is consequently in its normal operating state 81% of the time giving the desired stationary solution, $\mathbf{p}^* = [0.81\ 0\ 0\ 0\ 0.14\ 0.04\ 0.01]$.

Now, the information to start defining the probability matrix $\mathbf{P}$ is available. There are 19 unknown transition probabilities in the complete system as defined in Figure 4.16. However, the number of unknowns can be reduced in this example. As states 2–4 are not considered, transition probabilities p_{12}, p_{13} and p_{14} must be set to zero (meaning the probability to go to states 2–4 is zero), p_{22}, p_{33} and p_{44} should be set to one and p_{27}, p_{37} and p_{41} to zero as the sum of each row of $\mathbf{P}$ must equal one (a requirement for all

Markov chain systems). Now, 10 unknowns remain. The information obtained from the characterisation of the states is now used. As the sensor's average duration in states 5, 6 and 7 are known, the transition probabilities can be assigned to p_{55}, p_{66} and p_{77}, which in turn will give p_{51} and p_{67} as the sum of each row must equal one. The transition probability for p_{55} is simply 1 minus the inverse of the average duration in that state as it describes the probability to stay in the state: $p_{55} = 1 - 1/251 = 250/251$. The value of p_{55} means that the sensor will remain in state 5 for an average of 2.6 days once it has entered the state (based on a 15 minutes time step), which is what was concluded from the data series above. Similarly, $p_{66} = 353/354$ and $p_{77} = 3/4$ are the average probabilities to remain in states 6 and 7, respectively, once those states have been entered.

Now, the number of unknowns has been reduced to five: p_{11}, p_{15}, p_{17}, p_{71} and p_{76}. To determine the unique values for those parameters the information from the $\mathbf{p}^*$ vector is used together with the row sum condition = 1. In steady state, the expression $\mathbf{p}(k+1) = \mathbf{p}(k) \cdot \mathbf{P}$ must hold, where both $\mathbf{p}(k+1)$ and $\mathbf{p}(k)$ equal $\mathbf{p}^*$. The resulting linear equation system for this example is:

$$\begin{cases} 0.81 = 0.81 \cdot p_{11} + 0.14 \cdot 0.004 + 0.01 \cdot p_{71} \\ 0.14 = 0.81 \cdot p_{15} + 0.14 \cdot 0.996 \\ 0.04 = 0.04 \cdot 0.9972 + 0.01 \cdot p_{76} \\ 0.01 = 0.81 \cdot p_{17} + 0.04 \cdot 0.0028 + 0.01 \cdot 0.75 \\ p_{11} + p_{15} + p_{17} = 1 \\ p_{71} + p_{76} = 0.25 \end{cases}$$

From here the remaining parameters of the $\mathbf{P}$ matrix are easily calculated. The complete probability matrix becomes (rounded values, all empty matrix elements = 0):

$$\mathbf{P} = \begin{bmatrix} 0.9964 & 0 & 0 & 0 & 0.0007 & & 0.0029 \\ & 1.0 & & & & & 0 \\ & & 1.0 & & & & 0 \\ 0 & & & 1.0 & & & \\ 0.0040 & & & & 0.9960 & & \\ & & & & & 0.9972 & 0.0028 \\ 0.2387 & & & & & 0.0113 & 0.7500 \end{bmatrix}$$

Note that although the sensor is in the fully functional state 81% of the time, the transition probability to stay fully functional is 99.64% (p_{11}). This is perhaps somewhat surprising but follows from the high probabilities that the sensor will remain in states 5 and 6 for long periods of time when any of those states have been activated. It is important to note that no attempt has been made to mimic real measurements in terms of mean value, diurnal variation, or other. The focus here has been on the quality aspects of the measurements (i.e., faults and failures). In Figure 4.18, the simulated sensor is shown and illustrates that the modelled sensor displays similar behaviour in terms of faults and calibrations to the real measurements on which it is based (Figure 4.17).

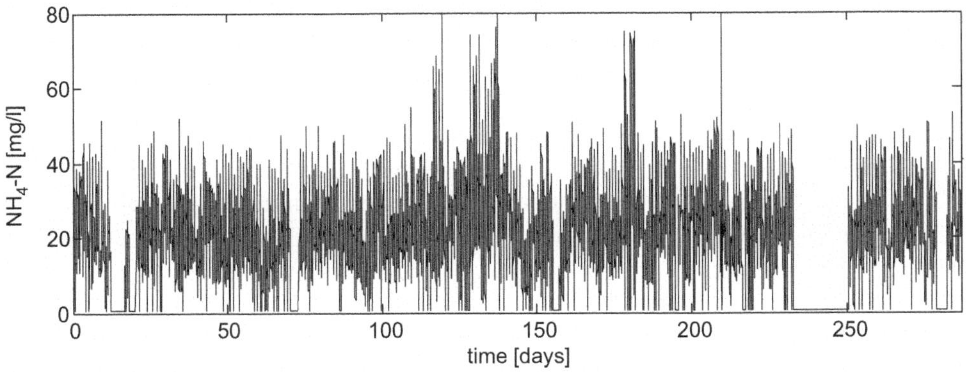

Figure 4.18 The simulated ammonia measurements with generated faults.

When using the complete sensor fault model (i.e., when states 2, 3 and 4 are also active) the above principle is still valid, however the calculation of the final eight unknown parameters is somewhat more complicated. Below, the general expression is provided for the complete case presented in Figure 4.16:

$$\begin{cases} p_{11} = ((p_{77}-1)\cdot p_7^* + (1-p_{66})\cdot p_6^* + (p_{55}-1)\cdot p_5^* + (p_{44}-1)\cdot p_4^* + p_1^*)/p_1^* \\ p_{12} = ((1-p_{22})\cdot p_2^*)/p_1^* \\ p_{13} = ((1-p_{33})\cdot p_3^*)/p_1^* \\ p_{14} = ((1-p_{44})\cdot p_4^*)/p_1^* \\ p_{15} = ((1-p_{55})\cdot p_5^*)/p_1^* \\ p_{17} = ((1-p_{77})\cdot p_7^* - p_{67}\cdot p_6^* - p_{37}\cdot p_3^* - p_{27}\cdot p_2^*)/p_1^* \\ p_{71} = ((1-p_{77})\cdot p_7^* + (p_{66}-1)\cdot p_6^*)/p_7^* \\ p_{76} = ((1-p_{66})\cdot p_6^*)/p_7^* \end{cases}$$

The same principle based on Markov chains is also used to model faults and failures of actuators in the BSMs. However, for actuators only five states are considered. More details can be found in Technical Report No. 2.

4.4 INHIBITION AND TOXICITY

Included in this benchmark model development is the ability to model inhibition and toxicity. This model addition addresses three different aspects of inhibition and toxicity:

- Effect on the biological processes;
- Physical propagation of inhibitory or toxic substances through the plant;
- Stochastic behaviour with respect to occurrence and magnitude of inhibition and toxicity events.

4.4.1 Biological processes

The inclusion of inhibition and toxicity has been achieved through the introduction of two pseudo states, S_{tox}, which represents inhibitory or toxic substances in a soluble form and X_{tox}, which represents substances in a particulate form or substances that are believed to be adsorbed onto other particulate matter in the process. It is assumed that S_{tox} and X_{tox} do not contribute to COD or TSS concentrations. The biological effect of the states is based on the combined concentration of S_{tox} and X_{tox}.

Inhibition is defined as the reduction in growth rates and toxicity is defined as the combined effect of reduced growth rates and increased decay rates. Autotrophic growth is assumed to be more sensitive to inhibition and is affected before heterotrophic growth. However, both types of biomass are equally susceptible to toxicity. The effect of the state concentrations on the growth and decay rates can be seen in Figure 4.19. The terms μ_{HT}, μ_{AT}, b_{HT} and b_{AT} represent temperature dependent growth and decay rates as defined in Section 4.6.

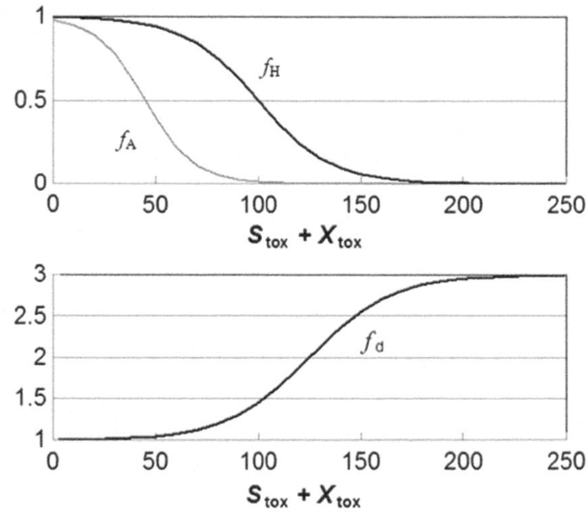

Figure 4.19 Illustrative effect of the combined concentration on growth and decay rates (f_H and f_A are the growth factors for heterotrophic and autotrophic growth rates, respectively, and f_d is the decay rate factor).

$$f_H = \frac{K_{tox,H}^{n_H}}{K_{tox,H}^{n_H} + (S_{tox} + X_{tox})^{n_H}}$$

$$f_A = f_H \cdot \frac{K_{tox,A}^{n_A}}{K_{tox,A}^{n_A} + (S_{tox} + X_{tox})^{n_A}}$$

$$f_d = 1.0 + 2.0 \cdot \frac{(S_{tox} + X_{tox})^{n_d}}{K_{tox,d}^{n_d} + (S_{tox} + X_{tox})^{n_d}}$$

$$\rho_1 = f_H \cdot \mu_{HT}\left(\frac{S_S}{K_S + S_S}\right)\left(\frac{S_O}{K_{O,H} + S_O}\right)X_{B,H}$$

$$\rho_2 = f_H \cdot \mu_{HT}\left(\frac{S_S}{K_S + S_S}\right)\left(\frac{K_{O,H}}{K_{O,H} + S_O}\right)\left(\frac{S_{NO}}{K_{NO} + S_{NO}}\right)\eta_g X_{B,H}$$

$$\rho_3 = f_A \cdot \mu_{AT}\left(\frac{S_{NH}}{K_{NH} + S_{NH}}\right)\left(\frac{S_O}{K_{O,A} + S_O}\right)X_{B,A}$$

$$\rho_4 = f_d \cdot b_{HT} X_{B,H}$$

$$\rho_5 = f_d \cdot b_{AT} X_{B,A}$$

4.4.2 Physical processes

A soluble and particulate state approach allows for temporal differentiation, that is, the soluble state will be retained in the system for a much shorter time compared to the particulate state (Figure 4.20). S_{tox} follows the transport dynamics of other solubles, whereas X_{tox} follows the mass balances of the particulate matter. S_{tox} is thus only affected by dilution and its retention time will be relatively short with an exponential decline in concentration after reaching a maximum. On the other hand, as S_{tox} and X_{tox} do not decay in the model, the retention time of X_{tox} is a function of the sludge retention time. Thus, S_{tox} and X_{tox} will have different effects on the plant performance and will challenge both control strategy development as well as process monitoring algorithm development.

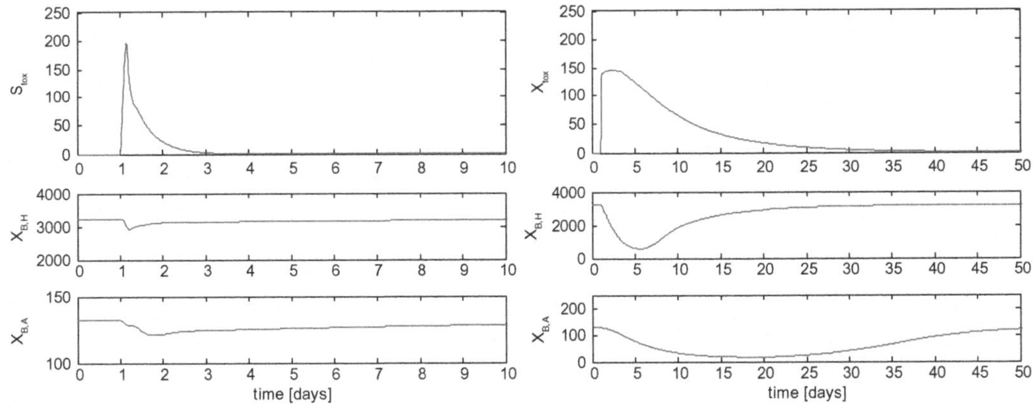

Figure 4.20 Concentration of S_{tox} (left) and X_{tox} (right) and their effects on heterotrophic and autotrophic biomass as a function of time after entering the plant as a pulse injection. Note the different time scales.

4.4.3 Modelling inhibitory/toxic substances

As with the sensor and actuator states, a Markov chain model was used to generate a basic 'seed' for S_{tox} and X_{tox}. The general Markov model that controls each component (S_{tox} and X_{tox}, i.e., two separate models) is depicted in Figure 4.21. As can be seen, the approach consists of three states (s_i) where s_1 represents no disturbance (no discharge of inhibitory/toxic substances), s_2 represents an inhibitory discharge (lower amounts of discharged inhibitory/toxic substances) and s_3 represents a toxic discharge (higher amounts of discharged inhibitory/toxic substances). As can be seen from the figure, it is assumed that it is possible to go from s_1 to s_2 or s_3 but from s_2 and s_3 it is only possible to go back to s_1. Translated into the manifestation of the disturbance, this means that when a discharge on the inhibitory level has occurred, a discharge on the toxic level can only occur after the state has returned to normal (no discharge). This mechanism is added to avoid a complete collapse of the biological processes during simulations.

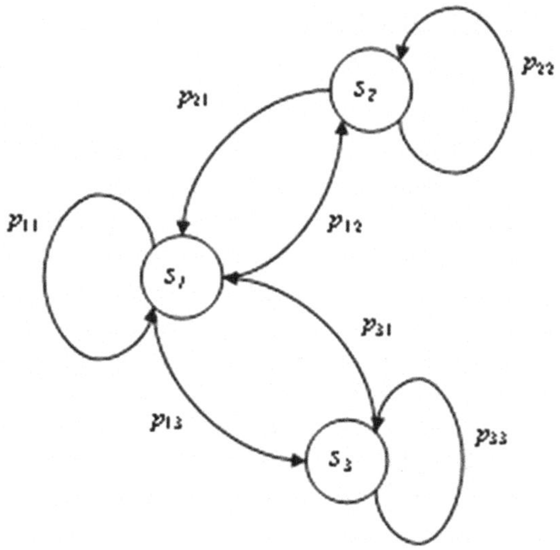

Figure 4.21 The three-state Markov chain model that governs the occurrence and duration of inhibitory/toxic discharges. s_1: no discharge, s_2: inhibitory discharge, s_3: toxic discharge.

The average duration of each discharge (i.e., the time spent in state s_2 or s_3) is assumed, for BSM1_LT, to be three hours. As the duration of a discharge is a stochastic variable, the flux per discharge event will vary significantly from event to event, resulting in a varying influent concentration. For benchmarking purposes, so as to ensure comparable results, this model was used to generate a 'seed' file (downloadable from www.benchmarkwwtp.org) that defines the toxic influent discharges over 609 days with a sampling time of 15 minutes. This 'seed' file feeds into the sewer model (Section 4.1) of the influent generator, meaning that the hydraulic conditions at the time of the event influence the incoming concentration. The influent concentrations of S_{tox} and X_{tox} in the standard BSM1_LT influent file and in the last reactor of the plant are shown in Figure 4.22.

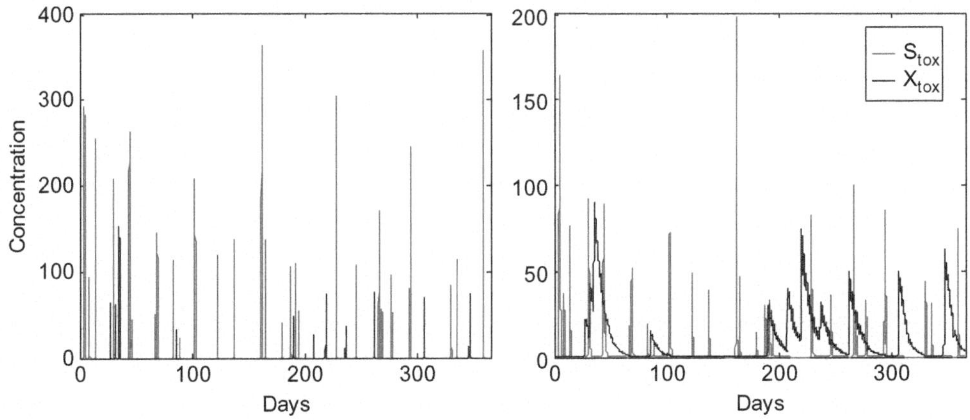

Figure 4.22 The Markov chain model generated concentration of S_{tox} and X_{tox} in the influent (left) and in the last biological reactor (reactor 5) in the plant (right).

4.5 RISK ASSESSMENT MODELLING

4.5.1 Concept

Progress in mechanistic modelling has improved activated sludge process descriptions, but common operational issues related to microbial population dynamics are still not well understood. Many of these problems relate to population imbalances between filamentous and floc-forming organisms leading to filamentous bulking, filamentous foaming and deflocculation (Wanner, 1998; Jenkins *et al.* 2003; Martins *et al.* 2004a). It is not uncommon to combine heuristic knowledge from the operators with specific data streams to identify these problems. However, the predictive power of this approach is limited as these data tend to appear only after the onset of the problem. The presence of heuristics and qualitative knowledge on complex phenomena such as filamentous bulking, foaming and rising sludge stands in sharp contrast with the lack of basic mechanistic knowledge on the population dynamics of the microorganisms causing these phenomena. As a consequence, although attempts have been made to explain the development of filamentous bacteria by means of mathematical modelling (Chen & Beck, 1993; Hug *et al.* 2006), none of these attempts has led to a general and experimentally validated model (Martins *et al.* 2004b). As an alternative, the benchmark system applies a risk assessment model, which integrates empirical knowledge with the mechanisms of standard deterministic models to infer solids separation problems of microbiological origin (Comas *et al.* 2008). The intention is to propose a risk assessment model for microbiology-related problems using only information available in the simulation outputs, either directly or after simple data processing. The reader is referred to the published literature and Technical Reports No. 11 and 12 for a full description of all three risk models (Comas *et al.* 2003, 2008). A general overview of the proposed approach is illustrated here using filamentous bulking.

4.5.2 Application to filamentous bulking

4.5.2.1 Decision tree

Initially, the knowledge related to the risk of filamentous bulking proliferation was synthesised into a decision tree (Figure 4.23). Each branch of the tree evaluates one assumed cause: (a) low dissolved oxygen concentration (left); (b) nutrient deficiency (middle); and (c) low F/M ratio or substrate limiting conditions (right). It has been reported that septic conditions or low pH in the influent can contribute to filamentous bulking, but the BSM models do not include sulphur or activated sludge pH modelling, so these conditions were not considered.

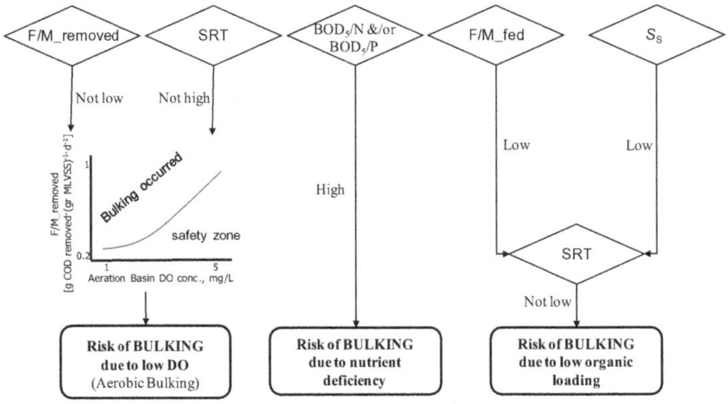

Figure 4.23 Decision tree developed to evaluate the risk of filamentous bulking (Comas *et al.* 2008).

The decision tree shows that up to seven variables can be used as indicators to assess the risk of filamentous bulking: SRT, DO, F/M removed (measured as g of removed COD per g biomass per day), F/M_fed (measured as g BOD_5 supplied per g biomass per day), BOD_5/N, BOD_5/P and S_S.

4.5.2.2 Modelling approach

The mathematical representation of the decision tree is captured using the principles of fuzzy decision theory (Figure 4.24).

The risk estimation involves three main steps:

(i) fuzzification;
(ii) fuzzy inference of the risk through a Mamdani approach (Mamdani & Assilan, 1975);
(iii) defuzzification of the output variable.

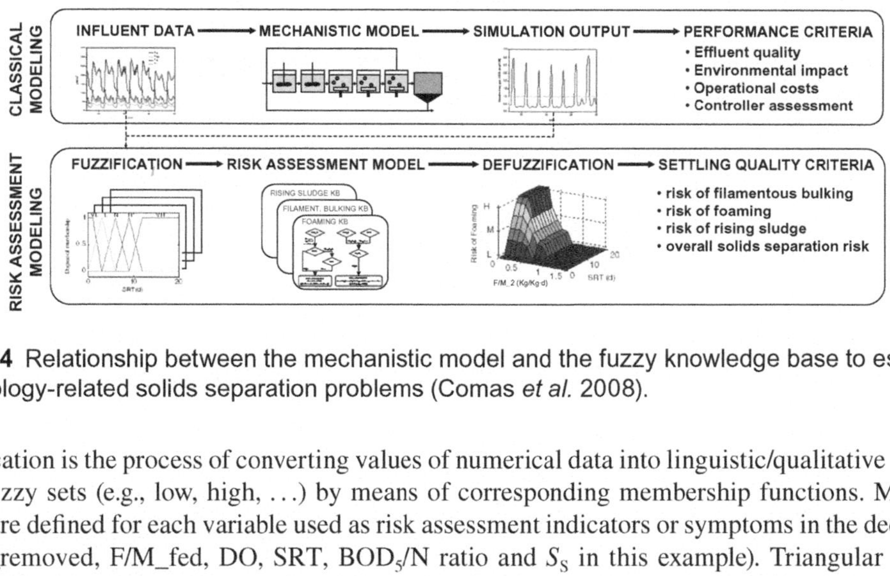

Figure 4.24 Relationship between the mechanistic model and the fuzzy knowledge base to estimate risk of microbiology-related solids separation problems (Comas *et al.* 2008).

Fuzzification is the process of converting values of numerical data into linguistic/qualitative descriptors or input fuzzy sets (e.g., low, high, ...) by means of corresponding membership functions. Membership functions are defined for each variable used as risk assessment indicators or symptoms in the decision trees (i.e., F/M_removed, F/M_fed, DO, SRT, BOD_5/N ratio and S_S in this example). Triangular or pseudo-trapezoidal functions are used to define the membership functions. Figure 4.25 illustrates examples of membership functions used in this approach. Table 4.8 lists a couple of membership function examples with their corresponding shapes and ranks. The limits of these membership functions as well as their degree of overlapping can be customised by the user according to the configuration and characteristics of the simulated activated sludge plant.

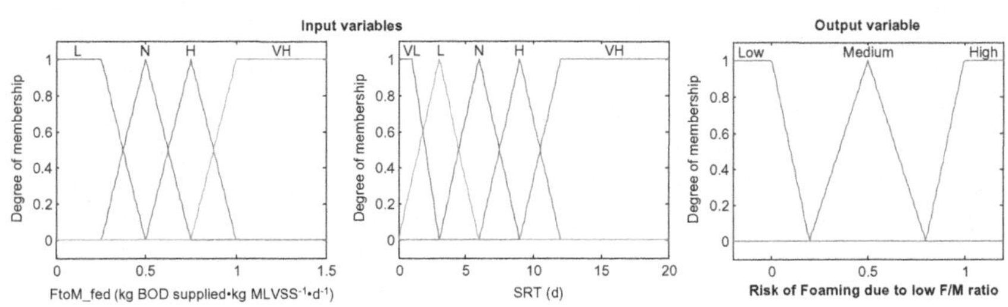

Figure 4.25 Example of membership functions (Comas *et al.* 2008).

Table 4.8 F/M_fed and SRT membership functions as defined in the BSM model.

Variable		Very low	Low	Normal	High	Very high
F/M_fed	Shape		trapezoidal	triangular	triangular	trapezoidal
($g \cdot g^{-1} \cdot d^{-1}$)	rank	–	[–0.0536 –0.0536 0.25 0.5]	[0.25 0.5 0.75]	[0.5 0.75 1]	[0.75 1 1.51 1.57]
SRT (d)	Shape	trapezoidal	triangular	triangular	triangular	trapezoidal
	rank	[–7.2 –0.8 1 3]	[0 3 6]	[3 6 9]	[6 9 12]	[9 12 20.29 23.4]

Fuzzy inference uses a Mamdani approach (Mamdani & Assilan, 1975) to generate a fuzzy output from the corresponding input fuzzy sets based on the fuzzy rule base. The rule base for bulking in the risk assessment model is shown in Table 4.9. The fuzzy rules for the BSM were derived from existing empirical knowledge and the literature.

Table 4.9 Relative risk of filamentous bulking as synthesised from bulking knowledge.

Bulking due to low DO

		$S_{O,3}$				
		VL	L	N	H	VH
F/M_removed	L	Low	Low	Low	Low	Low
	N	High	Medium	Low	Low	Low
	H	High	High	Medium	Low	Low
	VH	High	High	High	Medium	Low

Bulking due to low organic loading

		SRT				
		VL	L	N	H	VH
$S_{S,1}$	L	Low	Low	Medium	High	High
	N	Low	Low	Low	Low	Low
	H	Low	Low	Low	Low	Low
F/M_fed	L	Low	Low	High	High	High
	N	Low	Low	Low	Medium	Medium
	H	Low	Low	Low	Low	Low
	VH	Low	Low	Low	Low	Low

Bulking due to nutrient deficiency

$BOD_{5,in}/N_{in}$ or $BOD_{5,in}/P_{in}$	L	Low
	N	Low
	H	High

Note: VL: very low, L: low, N: normal, H: high; VH: very high.

Defuzzification of the output variable involves the translation of the linguistic fuzzy output into a numerical value. Membership functions for all three output variables (risk of filamentous bulking, risk of foaming and risk of rising sludge) were defined by means of the Center of Gravity (COG) method (Fiter *et al.* 2005).

The estimation of the risk is calculated continuously during the simulated time period with a new value calculated at every time step. For long simulation periods, the results of the risk assessment model are smoothed by means of an exponential filter. The exponential filter has a time constant related to the dynamics of each specific problem (e.g., three days for filamentous bulking). A further benefit of the exponential filter is that it facilitates the visualisation and interpretation of the results.

Work towards integrating the risk model into a running simulation has been carried out by Flores-Alsina *et al.* (2009), making it possible to link the model output to dynamic settling behaviour; however, it is important to understand that, as defined, the risk model runs in isolation. That is, although the purpose of this assessment is to provide a relative estimate of the likelihood of settling problems, the output is not linked to the settling model explicitly and the model does not predict settling problems just because the conditions indicate a risk at a particular instant. It must be assumed that the occurrence of a problem results after a high risk has been experienced for a prolonged and sustained period of time.

4.5.2.3 Temperature effect

The literature suggests that bulking and foaming problems due to the abundance of *M. parvicella* follow a typical seasonal pattern with massive growth of this microorganism occurring in the winter and early spring (Rosseti *et al.* 2005; Spering *et al.* 2008). This observation has been attributed to low temperature. So, in order to include this temperature effect on the risk assessment model, two of the individual risks, (i.e., the risk of foaming due to low F/M ratio and the risk of bulking due to low organic loading which are both related to *M. parvicella*), are multiplied by a temperature-dependent factor. The intended result of the factor is to increase the risk at low temperatures and decrease the risk at high temperatures. More details can be found in Comas *et al.* (2008).

4.5.2.4 Risk assessment outcomes

The results from the risk assessment are reported and quantified in four different ways for each of the defined settling problems. Note that the absolute values for different risks sometimes appear to be quite high. However, they should only be used for relative comparisons. For the integrated risks of bulking and foaming and for the overall risk index reporting is as follows:

- a time series plot showing the evolution of the calculated risk during the evaluation period. Standardisation of the plot means that 0 indicates no risk while 1 indicates the highest possible risk;
- the percentage of time the plant is at risk of a specific settling problem;
- the percentage of time the calculated risk is severe (an arbitrary value ≥ 0.8 is used in the BSM for defining a severe problem); and,
- the longest time interval that the plant is exposed to an uninterrupted severe risk of experiencing a specific settling problem.

4.6 TEMPERATURE

Temperature is known to affect the wastewater treatment process mainly through its impact on chemical and biological reaction rates. The loss of nitrification in winter has been attributed to temperature and temperature is known to affect oxygen transfer. More subtle effects such as population shifts that may give rise to increased risk of bulking in certain seasons are also receiving increased attention.

Temperature is included as an additional input variable in the influent model (Gernaey *et al.* 2006; 2011) and modelled by two types of phenomena. The seasonal temperature variations are implemented as a sinusoidal wave with a period of 364 days, an average value of 15°C and an amplitude of 5°C. The minimum influent temperature is reached in late January, the maximum influent temperature is reached in late July. As a second phenomenon, diurnal influent temperature changes are also included, and are implemented as a sinusoidal wave with a period of 1 day and an amplitude of 0.5°C. The diurnal influent temperature variations result in the lowest influent temperature in the early morning and highest in the late afternoon. No temperature effects are associated with rain events or snow melt.

Temperature is propagated through the model using a simple flux based model. That is, in each completely mixed reactor with a defined volume (except the digester, which has a fixed temperature of 35°C), temperature dynamics are modelled using a first-order system based on the 'heat' content ($T \cdot V$) in the tank and the heat content of the incoming water flux. Temperature at junction points is handled in the same way with the heat content of the streams 'flow-weighted' to give an output flow temperature.

As mentioned previously, some changes in biological activity are attributed to changes in temperature. So, to account for this, several of the model parameters are assumed to be temperature dependent (see Appendix A). The effects of temperature are calculated as described by Henze *et al.* (1987) using the Arrhenius relationship. In the default case, the following equation can be used to adjust temperature sensitive parameters

$$p_T = p_{15} \cdot \exp\left(\left(ln\left(\frac{p_{15,b}}{p_{10,b}}\right)/5\right) \cdot (T - 15)\right)$$

where p_T is the parameter value at temperature T, $p_{15,\mathrm{b}}$ is the benchmark-defined parameter value at 15°C and $p_{10,\mathrm{b}}$ is the benchmark-defined parameter value at 10°C and p_{15} is the simulation-specific parameter value at 15°C. Under normal circumstances, $p_{15,\mathrm{b}}$ and p_{15} will be the same quantity. However, to account for the possibility that a user may wish to change the value of p_{15} in a specific simulation, a distinction in these parameters has been included in the equation. The distinction ensures that the temperature effect is properly applied irrespective of the 15°C value of the parameter entered by a user.

Temperature also affects aeration efficiency and thus energy consumption through its effects on the mass transfer coefficient K_La and the oxygen saturation concentration, S_0^{sat}.

The solubility of oxygen is dependent on temperature, increasing with decreasing temperature. The ADM Task Group recommended general use of the constant enthalpy in form of the van't Hoff equation to describe temperature dependency of equilibrium constants (Batstone *et al.* 2002). This recommendation follows from the fact that the van't Hoff equation is a fundamental equation that uses a single well-known and easily measurable constant (enthalpy of reaction). For similar reasons, this Task Group agreed to use the C_p form of the van't Hoff equation (Lide *et al.* 2004) for the oxygen saturation concentration (valid in the range 273.15 K to 348.15 K):

$$S_0^{\mathrm{sat}}(T) = \frac{8}{10.50237016} \cdot 6791.5 \cdot K(T_{\mathrm{K}})$$

with

$$K(T_{\mathrm{K}}) = 56.12\, e^{-66.7354 + \frac{87.4755}{T^*} + 24.4526 \cdot ln(T^*)} \quad \text{and} \quad T^* = T_{\mathrm{K}} / 100$$

The term 8/10.50237016 ensures that the S_0^{sat} value at 15°C is exactly 8 g · m^{-3} and this maintains consistency with BSM1.

The oxygen transfer coefficient, K_La, is also dependent on temperature. ASCE (1993) presents the generally accepted dependency between the oxygen transfer coefficient K_La and temperature:

$K_La(T) = 1.024^{(T-15)} \cdot K_La_{15}$ with K_La in d^{-1} and T in °C.

This equation must be interpreted as follows: for a given energy input (or equivalently, a certain gas flow rate), which would yield a K_La of 240 d^{-1} at 15°C, the oxygen transfer efficiency drops to approximately 213 d^{-1} at 10°C, and increases to 270 d^{-1} when the temperature in the mixed liquor increases to 20°C.

It was decided that the aeration control handle should continue to be based on $K_La(15)$ with the modelled $K_La(T)$ being calculated based on the current model temperature. What this means is that to reach a certain oxygen concentration a higher $K_La(15)$ will be required if the reactor temperature is low and a lower value when the temperature is high. This leads to the known phenomenon that more energy will be needed for aeration at low temperatures compared to high temperatures. Fortunately, the variation in the oxygen saturation concentration with temperature compensates for this reduced mass transfer efficiency, so that overall, the energy needed to reach a certain oxygen concentration is not affected all that much by temperature (i.e., thanks to the increased driving force at low temperatures, a lower $K_La(15)$ will do the job).

Chapter 5

Benchmarking of control strategies

K. V. Gernaey, J. B. Copp, U. Jeppsson, I. Nopens, M.-N. Pons and P. A. Vanrolleghem

The main purpose of the benchmark models is the simulation-based comparison of control and process monitoring strategies. However, in practice the major part of this Scientific and Technical Report is focused on the modelling aspects of wastewater treatment plants. This illustrates that careful selection of each unit process model, its implementation and ring-testing, the fine tuning of its parameters and finally its approval as part of the benchmark models has been a valuable but also time-consuming exercise. Here, the Task Group would like to add explicitly that no realistic evaluation of control strategies would have been possible without these modelling efforts. Thus, in order to highlight the fundamental purpose of the benchmark models, this chapter focuses on benchmarking of control strategies. More specifically, certain aspects of controllers for BSM1, BSM1_LT and BSM2 will be highlighted.

5.1 BSM1 AND BSM1_LT CONTROLLERS

5.1.1 Default BSM1 control strategy

Default controllers are proposed for BSM1, such that the closed-loop simulation and the implementation of the evaluation criteria (Chapter 6) can be verified using the default strategy before the user implements his/her own control strategy. In this way the user is given an additional opportunity – in addition to the steady state and open-loop dynamic simulation results – to ensure that their BSM1 implementation is functioning properly. The primary control objectives for the default strategy in BSM1 are: (i) to maintain the NO_3-N concentration in the second anoxic tank at a predetermined set point value (1 g N · m^{-3}); and, (ii) to maintain the dissolved oxygen concentration in the fifth tank, that is, the last aeration tank, at a predetermined set point value (2 g O_2 · m^{-3}). Both default controllers are Proportional-Integral (PI) controllers. It should be emphasised that this defined control strategy is not the best strategy available. It is only provided as a test case for the user.

In this case, the NO_3-N sensor in the second anoxic compartment is a class B_0 sensor with a measurement range from 0 to 20 g N · m^{-3}. An explanation of the BSM sensor classes can be found in Chapter 4. The measurement noise is equal to 0.5 g N · m^{-3}. The manipulated variable is the internal recycle flow rate from the last aerated tank back to the first tank (Q_{int}).

For the dissolved oxygen control in the last aerated compartment, the probe is assumed to be a class A sensor with a measurement range from 0 to 10 g $O_2 \cdot m^{-3}$ and measurement noise of 0.25 g $O_2 \cdot m^{-3}$. The manipulated variable is the oxygen transfer coefficient for the last tank, K_La_5.

To increase the degree of realism in the simulations, constraints on the manipulated variables are applied. The internal recirculation flow (Q_{int}) is constrained to between 0 and 5 times the average dry weather influent flow rate to the plant (18 446 $m^3 \cdot d^{-1}$). The external recycle flow rate (Q_r) is held constant at the average dry weather influent flow rate to the plant. For the oxygen transfer in tank 5, it is assumed that K_La can only vary from 0 to 360 d^{-1}. Furthermore, whenever the K_La in an aerated tank is below 20 d^{-1}, it is assumed that mechanical mixing is needed in order to keep the activated sludge in suspension.

5.1.2 Other BSM1 control handles

All available control handles, except aeration (see below), are considered to be ideal. In the closed-loop test case, only two control handles are used: the internal recirculation flow rate (Q_{int}) and the oxygen transfer rate in tank number 5 (K_La_5). However, the following control handles are available for the implementation of user-defined control strategies in BSM1:

- return sludge flow rate (Q_r);
- anoxic/aerobic volume – all five biological tanks are assumed to be equipped with both aerators and mechanical mixing devices, that is, in a discrete fashion the volumes for anoxic and aerobic behaviour can be modified;
- aeration intensity individually for each tank (K_La_1, K_La_2, K_La_3, K_La_4, K_La_5), taking into account the dynamics of the aeration system;
- external carbon source flow rate (Q_{EC1}, Q_{EC2}, Q_{EC3}, Q_{EC4}, Q_{EC5}) where the carbon source is considered to consist of readily biodegradable substrate, that is, $COD_{EC} = S_S$;
- influent distribution by use of step feed (fractions of the influent flow to each of the five biological tanks: $f_{Qi1}, f_{Qi2}, f_{Qi3}, f_{Qi4}, f_{Qi5}$);
- distribution of internal flow recirculation (fractions of the internal recirculation flow to each of the five biological tanks: $f_{Qint1}, f_{Qint2}, f_{Qint3}, f_{Qint4}, f_{Qint5}$) – in principle internal flow recirculation between any of the tanks is allowed;
- distribution of return sludge flow (fractions of the return sludge flow to each of the five biological tanks: $f_{Qr1}, f_{Qr2}, f_{Qr3}, f_{Qr4}, f_{Qr5}$).

The non-ideal aeration system (K_La_1- K_La_5) is defined with significant dynamics and a response time of $t_r = 4$ min (see Rieger *et al.* 2006). A second-order time delay function gives a reasonable model of this process. The time constant of each of the two identical first-order delays is $\tau = t_r/3.89 = 1.03$ min.

The above selection of control handles dramatically increases the flexibility of the defined BSM1 system. This number of control handles may not be realistic for a real plant but they have been included here in order to encourage all types of control strategy evaluations.

5.1.3 BSM1_LT control strategy

Many control actions at a wastewater treatment plant act on the process over longer time scales and can thus not be evaluated using BSM1 with its one-week evaluation period. For example, a short-term evaluation does not allow for realistic equipment (sensor/actuator) modelling, including failures, drift and maintenance, as the impact of these phenomena typically appear over longer time scales. In order to cope with these limitations of BSM1, BSM1_LT was proposed and developed (Rosen *et al.* 2004; Corominas *et al.* 2011).

BSM1_LT has a one-year performance evaluation period, and includes seasonal temperature variations as well as additional influent disturbances in the form of inhibitory and toxic compounds that influence the process rates in the treatment plant model. Control handles such as the waste sludge flow rate (Q_w) can thus be evaluated with BSM1_LT.

Apart from long-term control strategy evaluations, BSM1_LT also aims at the evaluation of monitoring strategies, such as detecting undesirable sensor or actuator faults in the plant. Process monitoring as such is not very useful if the monitoring results do not benefit the plant operation. But, BSM1_LT offers one additional feature that distinguishes it and that is the possibility to study fault-tolerant control. BSM1_LT offers the possibility to test the impact of an alarm in an on-line sensor or actuator. Similar to treatment plants in practice, once detected by a suitable monitoring algorithm, a control action aimed at compensating for the problem, either operator-based or automatic adjustment (i.e., fault-tolerant control) eventually will have to be performed. In both cases, the monitoring system will be expected to lead to more robust plant performance. Economic evaluation of the resulting closed-loop plant performance may be more relevant to practice than the evaluation of the monitoring system on its own.

5.2 BSM2 CONTROLLERS

5.2.1 Default BSM2 control strategy

BSM2 includes a primary clarifier, the activated sludge plant and sludge treatment. As a consequence, BSM2 is suited for simulation and comparison of so-called plant-wide control strategies. Similar to the BSM1, a default control strategy has been developed for BSM2 with the aim to demonstrate the potential for control actions. It should be emphasised that this proposed control strategy does not represent the best strategy available. The main purpose of the default strategy is to provide a simple example of how the benchmark can be used to develop novel plant-wide control strategies and to verify that the implementation works correctly.

The default closed-loop configuration for BSM2 consists of a PI dissolved oxygen (DO) controller that controls the DO in tank 4 (set point: 2 g $O_2 \cdot m^{-3}$) by manipulating K_La_3, K_La_4 and K_La_5 with K_La_5 set to half the value of K_La_3 and K_La_4. Note that the controller parameter values are not reported explicitly as they differ between platforms and depend on the controller implementation (as was the case for BSM1 as well). As in the open-loop case, an external carbon source is fed to the first anoxic reactor at a constant flow rate of 2 $m^3 \cdot d^{-1}$ (COD concentration (as S_S): 400 000 $g \cdot m^{-3}$). Timer-based control of the wastage sludge flow rate (Q_w) is active. When the temperature of the influent wastewater is below 15°C, Q_w is set to 300 $m^3 \cdot d^{-1}$ (i.e., for t = 0–181 days and t = 364–454 days), and when the temperature is above 15°C, Q_w is set to 450 $m^3 \cdot d^{-1}$ (i.e., for the remaining time periods). This wastage flow control is fully modifiable in any user-defined control strategy.

Some results for the BSM2 default control strategy (CL0) are summarised in Table 5.1, along with results from the open-loop case (OL). It can be seen that the default strategy leads to a decrease in effluent S_{NH} compared to the open-loop case in part because of the added DO in the final aerated tank. At the same time, effluent S_{NO} increases considerably for the default control strategy because that additional DO is recycled to the unaerated part of the plant *via* the internal recycle and reduces denitrification. Both effects are the result of the additional aeration.

5.2.2 Testing other BSM2 control strategies

In order to further illustrate the use of the BSM2, two other simple control strategies are shown. The first control strategy (CL1) consists of a combination of two PI DO-controllers. The first one controls the DO in tank 4 (set point: 2 g $O_2 \cdot m^{-3}$) by manipulating K_La_3 and K_La_4 whilst the second one controls the DO in tank 5 (set point: 1 g $O_2 \cdot m^{-3}$) by manipulating K_La_5. This strategy is a refinement of the default closed-loop strategy and aims to reduce the excess oxygen supplied to tank 5 in the default closed-loop case. The results for this strategy are shown in Table 5.1. Due to better control of the DO in tank 5 and, hence, less DO recycled to the anoxic tank, denitrification is somewhat improved.

Table 5.1 Summary of simulation results for BSM2 open-loop (OL) and three closed-loop (CL) cases (average effluent concentrations).

	Unit	OL	CL0	CL1	CL2
Av $S_{NH,e}$	g N · m^{-3}	1.65	0.47	0.48	1.11
Av $S_{NO,e}$	g N · m^{-3}	7.47	11.05	10.40	7.85
Av TSS_e	g COD · m^{-3}	15.90	15.17	15.17	14.92
Av TN_e	g N · m^{-3}	11.20	13.53	12.89	10.94
Av $COD_{tot,e}$	g COD · m^{-3}	50.06	49.02	49.03	48.78

The second alternative control strategy (CL2) consists of a PI DO-controller and a cascade S_{NH}-DO controller. The DO control in tanks 3 and 4 is exactly the same as in case CL1. However, the DO in tank 5 is controlled using a PI S_{NH}-controller to dynamically define the DO set point within a defined range instead of using a fixed set point ($DO_{sp,min}$ = 0 g $O_2 \cdot m^{-3}$; $DO_{sp,max}$ = 3 g $O_2 \cdot m^{-3}$). For this case, the tank 5 set point for S_{NH} is 1.5 g N · m^{-3}. The PI DO-controller uses this set point to control K_La_5. And lastly, the carbon dosage is reduced to 1 $m^3 \cdot d^{-1}$.

This last strategy attempts to directly control effluent S_{NH} instead of DO. The previous cases (CL0, CL1) show that by controlling DO, nitrification is essentially complete and the effluent ammonia concentration is well below the effluent requirement. This means that more oxygen than necessary is being consumed, excess S_{NO} production is occurring and more external carbon may be required to keep the total nitrogen (TN) concentration below its limit. The CL2 strategy aims to limit this excess nitrification. The strategy drastically reduces the effluent S_{NO}, even with the lower carbon dosage. Furthermore, although the DO variations are wider compared to the other cases, the average oxygen consumption is also reduced significantly. This is a nice example of how a well-designed control strategy can improve plant performance and, at the same time, decrease costs and energy usage.

Chapter 6

Evaluation criteria

D. Vrecko, E. I. P. Volcke, U. Jeppsson, K. V. Gernaey, J. B. Copp and P. A. Vanrolleghem

The definition of the BSM simulation procedure discussed in the next chapter is meant to ensure that any data analysed by simulation benchmark users is generated in a similar manner. However, the results of these dynamic simulations lead to the question of how the huge amount of output data is to be evaluated. To aid the evaluation process, a series of quantities have been identified. These evaluation criteria provide a simple, accurate and objective means for comparison and have been developed specifically for comparing dynamic responses and the impact of different control strategies.

The evaluation criteria, as a whole, are a set of geographically independent measures that combine the output data into a small number of composite terms. The criteria discussed here update and extend the criteria defined for BSM1 (Copp, 2002; Pons *et al.* 2009), to include the new processes included in BSM2. These composite terms include, among others, a general effluent quality measure, an operational cost index, energy terms for pumping and aeration, cost for external carbon addition, biogas production and a measure of sludge production. The methods used to calculate these terms are outlined below.

To exemplify the evaluation criteria, a complete evaluation example of the open-loop BSM2 is presented in Appendix A. The open-loop BSM2 case is taken directly from Nopens *et al.* (2010) and was simulated exactly according to the description given there. The influent data file used is the one defined by Gernaey *et al.* (2005, 2006, Technical Report No. 8) and available for download at www.benchmarkwwtp.org.

6.1 EFFLUENT AND INFLUENT QUALITY INDICES

The effluent quality index (EQI, in kg pollution units $\cdot$ d^{-1}) is a weighted average sum of relevant effluent concentrations (Copp, 2002). The averaged values used for all evaluation criteria for all BSMs are based on dynamic output values recorded every 15 minutes and assume constant values between samples (zero-order hold). However, the evaluation varies slightly depending on the benchmark system being used. For example, with BSM1, the evaluation period (t_{obs}) is one week (t_{start} = 21 days, t_{end} = 28 days) but in the case of BSM2 and BSM1_LT, the evaluation period (t_{obs}) has been extended to 364 days (t_{start} = 245 days, t_{end} = 609 days). Secondly, whereas there is only one effluent in BSM1 and BSM1_LT (Figure 3.1), the effluent that is used to calculate the EQI for BSM2 (Figure 3.2) consists of the combination of the settler effluent, the effluent that bypassed the plant entirely and the effluent that bypassed the activated sludge system only (i.e., this last flow has passed through the primary settler). In BSM2, those three effluent sources are added together and the sum (Q_e in the equation below) is used for the evaluation of the effluent criteria:

$$\mathrm{EQI} = \frac{1}{t_{\mathrm{obs}} 1000} \int_{t_{\mathrm{start}}}^{t_{\mathrm{end}}} \Big(\beta_{\mathrm{SS}} \cdot \mathrm{TSS}_e(t) + \beta_{\mathrm{COD}} \cdot \mathrm{COD}_e(t) + \beta_{\mathrm{TKN}} \cdot \mathrm{TKN}(t) + \beta_{\mathrm{NO}} \cdot S_{\mathrm{NO},e}(t) + \beta_{\mathrm{BOD5}} \cdot \mathrm{BOD}_{5,e}(t) \Big) \cdot Q_e(t) \cdot dt$$

The definitions for the composite variables are similar for all the systems (Copp, 2002) and the weighting factors (β_x) for the different concentrations are based on Vanrolleghem *et al.* (1996) with the TKN and S_{NO} factors modified to promote ammonia removal. The values for the weighting factors are: $\beta_{\mathrm{SS}} = 2$, $\beta_{\mathrm{COD}} = 1$, $\beta_{\mathrm{TKN}} = 30$, $\beta_{\mathrm{NO}} = 10$ and $\beta_{\mathrm{BOD5}} = 2$.

$$\begin{aligned}
\mathrm{TSS} &= 0.75 \cdot (X_{\mathrm{S}} + X_{\mathrm{I}} + X_{\mathrm{B,H}} + X_{\mathrm{B,A}} + X_{\mathrm{P}}) \\
\mathrm{COD} &= S_{\mathrm{S}} + S_{\mathrm{I}} + X_{\mathrm{S}} + X_{\mathrm{I}} + X_{\mathrm{B,H}} + X_{\mathrm{B,A}} + X_{\mathrm{P}} \\
\mathrm{TKN} &= S_{\mathrm{NH}} + S_{\mathrm{ND}} + X_{\mathrm{ND}} + i_{\mathrm{XB}} \cdot (X_{\mathrm{B,H}} + X_{\mathrm{B,A}}) + i_{\mathrm{XP}} \cdot (X_{\mathrm{P}} + X_{\mathrm{I}})
\end{aligned}$$

It should be noted that ASM1 does not include a state variable for inorganic particulate matter so a completely realistic and dynamic TSS value cannot be modelled with ASM1. This recognised short-coming of ASM1 was compensated for by using a value of 0.75 to relate particulate COD to TSS. This approach is not ideal as it results in a perfect correlation between the inorganic particulate material and the total particulate COD content, but it is an attempt to acknowledge that inorganic material will exist in the system.

The definitions for TSS ($\mathrm{g \cdot m^{-3}}$), COD ($\mathrm{g\ COD \cdot m^{-3}}$) and TKN ($\mathrm{g\ N \cdot m^{-3}}$) are the same for all streams, but the definition for $\mathrm{BOD_5}$ ($\mathrm{g\ BOD_5 \cdot m^{-3}}$) is not. The benchmark system assumes that $\mathrm{BOD_5}$ in all streams is a function of the degradable organic state variables. However, different factors are applied to each stream depending on the source of that material. The assumption is made that the influent contains more readily biodegradable material and a factor of 0.65 is applied to the influent organic content. This contrasts the 0.25 factor used for the fully processed secondary effluent, which typically contains less readily biodegradable material and more biomass and slowly biodegradable organic matter. Of course all effluent that has bypassed the activated sludge system uses the influent factor, 0.65. As a consequence, the effluent $\mathrm{BOD_5}$ is calculated as:

$$\mathrm{BOD}_{5,\mathrm{e}} = (\mathrm{BOD}_{5,\mathrm{e,s}} \cdot Q_{\mathrm{e,s}} + \mathrm{BOD}_{5,\mathrm{e,bp}} \cdot Q_{\mathrm{e,bp}} + \mathrm{BOD}_{5,\mathrm{e,bas}} \cdot Q_{\mathrm{e,bas}}) / Q_{\mathrm{e}}$$

where,

$$\begin{aligned}
\mathrm{BOD}_{5,\mathrm{e,s}} &= 0.25 \cdot (S_{\mathrm{S,s}} + X_{\mathrm{S,s}} + (1 - f_{\mathrm{P}}') \cdot (X_{\mathrm{B,H,s}} + X_{\mathrm{B,A,s}})) \\
\mathrm{BOD}_{5,\mathrm{e,bp}} &= 0.65 \cdot (S_{\mathrm{S,bp}} + X_{\mathrm{S,bp}} + (1 - f_{\mathrm{P}}') \cdot (X_{\mathrm{B,H,bp}} + X_{\mathrm{B,A,bp}})) \\
\mathrm{BOD}_{5,\mathrm{e,bas}} &= 0.65 \cdot (S_{\mathrm{S,bas}} + X_{\mathrm{S,bas}} + (1 - f_{\mathrm{P}}') \cdot (X_{\mathrm{B,H,bas}} + X_{\mathrm{B,A,bas}}))
\end{aligned}$$

with the subscripts ‘s’, ‘bp’ and ‘bas’ referring to the settler effluent, effluent that bypassed the entire plant and the effluent that bypassed the activated sludge system, respectively.

Note: Thanks to the rigorous attention of one of the STR reviewers, the authors acknowledge that an error in these BOD equations has existed since they were first introduced nearly 20 years ago. Based on this review, the above equations now refer to an f_p' term (rather than f_p as previously used) and that constant should have a value of 0.20. The Task Group believes that this value is more consistent with modelling practice and differentiates this constant from the ASM1 model parameter, f_p. Two things should be further noted: i) although the parameter value change will have a minimal impact on the BOD calculations, it was deemed important to make this change; and, ii) all BOD results reported in the STR were calculated with the incorrect value, but going forward BOD values will be calculated using the corrected value of 0.20.

Besides the effluent quality, the influent quality can be calculated for reference purposes. The influent quality index (IQI, in kg pollution units $\cdot$ d^{-1}) is calculated in the same way as the effluent quality, as a weighted average sum of corresponding influent concentrations, with BOD_5 calculation using the 0.65 factor.

6.2 EFFLUENT CONCENTRATIONS

Specific components in the effluent are evaluated using three quantities including 95 percentile and violation reporting. Effluent limit violations are included so that the control strategies are compared not only based on their ability to meet a control objective, but also on their ability to provide effective treatment. The effluent violations are calculated from the output data generated at 15-minute intervals and are reported through two quantities: (i) number of violations; and, (ii) percentage of time the plant is in violation. Constraints with respect to five effluent components are defined and the percentage of time that the constraints are not met is to be reported. As well, the methodology for reporting the number of violations is defined. The same effluent limits are defined for all of the benchmark models:

- $N_{tot,e} < 18$ g N $\cdot$ m^{-3},
- $COD_e < 100$ g COD $\cdot$ m^{-3},
- $S_{NH,e} < 4$ g N $\cdot$ m^{-3},
- $TSS_e < 30$ g TSS $\cdot$ m^{-3} and,
- $BOD_{5,e} < 10$ g $BOD_5 \cdot m^{-3}$.

The above effluent limits were arbitrarily chosen and do not correspond to any effluent requirements in any known location. For the purposes of the benchmark, it should be noted that these limits are compared to the instantaneous model output generated at 15 minute intervals and are not related to daily, weekly, monthly, quarterly or yearly averages as might be the case with real regulations in a given location.

6.2.1 Ninety-five (95) percentiles

The 95 percentiles are defined as the effluent concentrations that are exceeded only 5% of the time. Percentile concentrations for effluent ammonia ($S_{NH,e95}$), effluent total nitrogen ($N_{tot,e95}$) and total suspended solids (TSS_{e95}) are to be reported as part of the evaluation criteria.

6.2.2 Number of violations

This quantity represents the number of times that the plant is in violation of the effluent limits (i.e., the number of times the plant effluent concentration increases above the effluent limit during the evaluation period). This measure does *not* reflect the length of time that the plant is in violation. An illustrative example from Copp *et al.* (2002) is given below.

6.2.3 Percentage of time plant is in violation

This quantity is a measure of the percentage of the time that the plant is in violation of the effluent limits. An illustrative example is given below.

Calculation Example:
(effluent constraint: 30 g·m^{-3})

time (h)	Effluent Suspended Solids (g SS·m^{-3})	
	Case 1	Case 2
0.00	25	25
0.25	34	34
0.50	26	37
0.75	37	26
1.00	22	22

Two hypothetical data sets for effluent suspended solids are shown in the table above and depicted in the figures below.

In case 1, the number of violations is 2. That is, the effluent suspended solids rose above the effluent constraint twice in this 1-hour period. During that hour, the effluent limit was in violation for 30 minutes (square wave concentrations are assumed, see figures); hence the percentage of time in violation is 50.

In case 2, the number of violations is 1. That is, the effluent suspended solids rose above the effluent constraint only once in this hour. The fact that it remained above the constraint for an extended period is irrelevant for this quantity. As in case 1, the effluent limit was in violation for 30 minutes so the percentage of time in violation is still 50 for case 2.

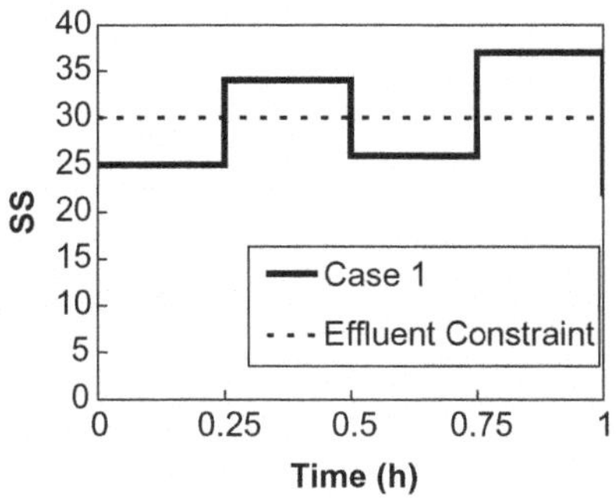

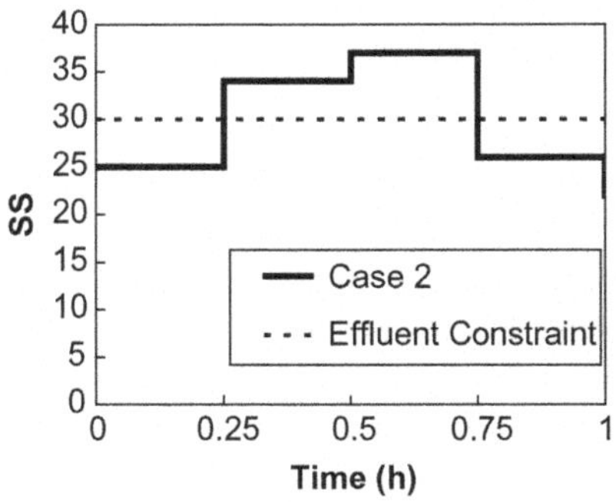

6.3 OPERATIONAL COST INDEX

The operational cost index (OCI) is calculated as a weighted sum of different costs:

$$\mathrm{OCI} = AE + PE + f_{sp} \cdot SP + f_{ec} \cdot EC + ME - f_{mp} \cdot MP + \max(0, HE - 7MP)$$

where, AE (kWh·d^{-1}) is aeration energy, PE (kWh·d^{-1}) is pumping energy, SP (kg SS·d^{-1}) is sludge production for disposal, EC is external carbon addition (kg COD·d^{-1}), ME (kWh·d^{-1}) is mixing energy, MP (kg $CH_4 \cdot d^{-1}$) is methane production and HE (kWh $\cdot d^{-1}$) is the heating energy needed to increase the temperature of the sludge in the anaerobic digester. The weighting factors are listed in Table 6.1.

Table 6.1 OCI weighting factors.

Factor	BSM1	BSM_LT	BSM2
f_{sp}	5	5	3
f_{EC}	3	3	3
f_{mp}	n/a	n/a	6

The value of f_{sp} was reduced for BSM2 to account for the sludge processing being carried out in that system and the expected sludge disposal cost savings as a result. The weight for EC, f_{EC}, is 3 and the value for f_{mp} is set to 6, which is consistent with the assumption that 43% of the theoretical energy content of the produced methane can be made available as electricity by the assumed gas motor. This also assumes that approximately 50% of the energy is used for heating the digester influent and 7% is lost.

The resulting OCI is dimensionless, but is a quantity that can be used to compare the relative costs between different control strategies.

6.3.1 Aeration energy

In the original BSM1 definition (Copp, 2002), the aeration energy (AE) calculation was based on the following equation with an output in units kWh $\cdot d^{-1}$:

$$AE = \frac{24}{t_{obs}} \int_{t_{start}}^{t_{end}} \sum_{i=1}^{i=5} (3.0248 \cdot 10^{-4} \cdot V_i \cdot K_La_{i,15}(t)^2 + 0.5882 \cdot 10^{-2} \cdot V_i \cdot K_La_{i,15}(t)) \cdot dt$$

where K_La_{15} was the K_La value calculated at a temperature of 15°C, K_La is expressed in h^{-1}, V_i is the volume of reactor i (m^3) and t_{obs} represents the total length of the evaluation period in days (i.e., $t_{end} - t_{start}$). Use of this equation demonstrated that, while the temperature effects on K_La and S_O^{sat} are considered properly, the calculated aeration energy costs tended to dominate the other operating cost quantities such as sludge production. As a result of these findings, a new approach for calculating the aeration cost was developed. This new approach was based on a typical engineering approach related to the SOTE (Standard Oxygen Transfer Efficiency). It was assumed in this approach that at 15°C and a dissolved oxygen concentration of 0 g $O_2 \cdot m^{-3}$, 1.8 kg O_2 can be transferred per kWh under process conditions. This allows AE as a function of K_La_{15} to be written down explicitly. The amount of oxygen transferred per unit time or more commonly referred to as the oxygen transfer rate (OTR), calculated in units kg $O_2 \cdot d^{-1}$ in a volume V, is given by:

$$\text{OTR} = V \cdot K_La_{15} \cdot (S_{O,15}^{sat} - 0)/1000$$

where, S_O^{sat} is expressed in g $O_2 \cdot m^{-3}$, V in m^3 and K_La_{15} in d^{-1}. For such an OTR, the energy required is given by:

$$AE = \frac{S_{O,15}^{sat}}{t_{obs} \cdot 1.8 \cdot 1000} \int_{t_{start}}^{t_{end}} V \cdot K_La_{15}(t)dt$$

Because the calculated AE is independent of the dissolved oxygen concentration (i.e., any K_La_{15} comes with an energy expenditure equal to the one that would lead to 1.8 kg $O_2 \cdot (kWh)^{-1}$ transferred when the DO was zero), the equation can be applied at all times, irrespective of the actual DO concentration. This leads to the new AE function:

$$AE = \frac{S_{O,15}^{sat}}{t_{obs} \cdot 1.8 \cdot 1000} \int_{t_{start}}^{t_{end}} \sum_{i=1}^{i=5} V_i \cdot K_La_{i,15}(t)dt$$

where, t_{obs} is the period of evaluation (d), V_i is the volume of reactor i (m^3), $K_La_{i,15}$ is the mass transfer coefficient in reactor i at 15°C (d^{-1}) and $S_{O,15}^{sat}$ is the saturated oxygen concentration at 15°C (i.e., 8 g $O_2 \cdot m^{-3}$). The modified AE calculation is now used in all BSMs.

6.3.2 Pumping energy

Pumping energy ($kWh \cdot d^{-1}$) is calculated as a weighted average sum of the various pumped flows in the system under study. In BSM1, this includes only the external recycle flow Q_r, the internal recycle flow Q_{int} and the waste sludge flow Q_w. In the original BSM1 definition a common weight of 0.04 $kWh \cdot m^{-3}$ was defined for all pumped flows, but this assumption was later modified to include separate weighting factors for each flow. In BSM2, the calculation also includes the primary clarifier underflow Q_{pu}, the thickener unit underflow Q_{tu} and the dewatering unit overflow flow Q_{do} to yield:

$$PE = \frac{1}{t_{obs}} \int_{t_{start}}^{t_{end}} \begin{pmatrix} f_{PE\_Q_{int}} \cdot Q_{int} + f_{PE\_Q_r} \cdot Q_r + f_{PE\_Q_w} \cdot Q_w + \\ f_{PE\_Q_{pu}} \cdot Q_{pu} + f_{PE\_Q_{tu}} \cdot Q_{tu} + f_{PE\_Q_{do}} \cdot Q_{do} \end{pmatrix} dt$$

where all flows are expressed in $m^3 \cdot d^{-1}$. Flows not included in this expression (between reactors for instance) are assumed to be gravitational with no energy requirement. For all BSMs, the following PE factors are suggested with only the first three applicable to BSM1 and BSM1_LT:

- f_{PE_Qint}: 0.004 $kWh \cdot m^{-3}$
- f_{PE_Qr}: 0.008 $kWh \cdot m^{-3}$
- f_{PE_Qw}: 0.050 $kWh \cdot m^{-3}$
- f_{PE_Qpu}: 0.075 $kWh \cdot m^{-3}$
- f_{PE_Qtu}: 0.060 $kWh \cdot m^{-3}$
- f_{PE_Qdo}: 0.004 $kWh \cdot m^{-3}$

6.3.3 Sludge production for disposal

Sludge production for disposal ($kg \cdot d^{-1}$) is calculated as the mass of solids accumulated in the plant or purposely removed from the process. In BSM1 and BSM1_LT this refers to the sludge removed in the waste activated sludge stream, whereas in BSM2 it is the solids removed from the plant as dewatered sludge. The following equation is used for the calculation:

$$SP = \frac{1}{t_{obs} \cdot 1000} \cdot \left(M_{TSS}(t_{end}) - M_{TSS}(t_{start}) + \int_{t_{start}}^{t_{end}} TSS_x(t) \cdot Q_x(t) \cdot dt \right)$$

where, $Q_x(t)$ ($m^3 \cdot d^{-1}$) is the sludge flow and TSS_x is the total solids concentration in the sludge flow stream (after dewatering in BSM2). M_{TSS} is defined as the sum of the total suspended solids mass (in kg) present in each of the individual unit processes. For BSM1 and BSM1_LT this includes the activated sludge reactors and the secondary clarifier. For BSM2 the primary clarifier, anaerobic digester and sludge storage tank are also included. M_{TSS} is calculated as follows:

$$M_{TSS}(t) = M_{TSS,as}(t) + M_{TSS,sc}(t) + M_{TSS,pc}(t) + M_{TSS,ad}(t) + M_{TSS,ss}(t)$$

with

$$M_{TSS,x}(t) = TSS_x(t) \cdot V_x$$

and, subscripts 'as', 'sc', 'pc', 'ad' and 'ss' referring to the activated sludge reactors, the secondary clarifier, the primary clarifier, the anaerobic digester and the sludge storage tank, respectively. It should be noted that the total solids concentration in the anaerobic digester (TSS_{ad}) is calculated using the ASM1 states generated in the output stream *after* the ADM1/ASM1 interface. This special case eliminates the need to define a new TSS calculation using ADM1 state variables and clarifies that the calculated TSS throughout the BSM2 layout is based on ASM1 state variables. The reason for this detailed calculation of sludge mass in the systems is to avoid the possibility of temporarily 'storing' sludge in the reactors during the evaluation period, thereby producing less sludge for disposal.

6.3.4 External carbon

External carbon addition ($kg\ COD \cdot d^{-1}$) is calculated as the total external carbon mass flow:

$$EC = \frac{COD_{EC}}{t_{obs} \cdot 1000} \int_{t_{start}}^{t_{end}} Q_{EC}(t) \cdot dt$$

where, COD_{EC} is the external carbon source concentration ($4 \cdot 10^5\ g\ COD \cdot m^{-3}$) and Q_{EC} ($m^3 \cdot d^{-1}$) is the sum of external carbon source flow rates into the biological reactors.

6.3.5 Mixing energy

Mixing energy ($kWh \cdot d^{-1}$) combines energy used for mixing the activated sludge tanks (ME_{as}) with the energy used for mixing the anaerobic digester (ME_{ad}):

$$ME = ME_{as} + ME_{ad}$$

Mixing is assumed to be required in a reactor if that reactor is unaerated or if the applied K_La_i in that reactor is smaller than 20 d^{-1}. Aeration is assumed to provide sufficient mixing when the K_La_i in a reactor is greater than 20 d^{-1}. The energy for activated sludge tank mixing is calculated using:

$$ME_{as} = \frac{24}{t_{obs}} \int_{t_{start}}^{t_{end}} \sum_{i=1}^{i=5} \begin{bmatrix} \text{for } t \text{ when } K_La_i(t) < 20\,d^{-1} \text{ then } f_{ME_{as}} \cdot V_i \\ \text{for } t \text{ when } K_La_i(t) \geq 20\,d^{-1} \text{ then } 0 \end{bmatrix} \cdot dt$$

where, V_i is the ith tank volume (m^3) and $f_{ME_{as}}$ is the mixing power consumption factor for the activated sludge tanks, which is set to 0.005 $kW \cdot m^{-3}$. It is assumed that the anaerobic digester is mixed constantly, so the mixing energy can be calculated simply as:

$$ME_{ad} = 24 \cdot f_{ME_{ad}} \cdot V_{ad}$$

where, V_{ad} is the liquid volume (m^3) of the anaerobic digester and $f_{ME_{ad}}$ is the mixing power factor for the anaerobic digester which is set to the same value as the activated sludge factor, that is, 0.005 $kW \cdot m^{-3}$. In the BSM1 and BSM1_LT systems ME_{ad} does not exist.

6.3.6 Methane production

Methane production (kg $CH_4 \cdot d^{-1}$) in the anaerobic digester represents an economic benefit and is included in the OCI as a negative value. An average value of the quantity of methane produced per day is derived (based on the general gas law) using:

$$MP = \frac{P_{atm} \cdot 16}{t_{obs} \cdot R \cdot T_{op}} \int_{t_{start}}^{t_{end}} \frac{1}{p_{gas,tot}(t)} \cdot p_{gas,CH4}(t) \cdot Q_{gas}(t) \cdot dt$$

where, $p_{gas,CH4}$ (bar) is the partial pressure of the produced methane gas in the headspace, R is the universal gas law constant ($8.3145 \cdot 10^{-2}$ $bar \cdot m^3 \cdot kmol^{-1} \cdot K^{-1}$), T_{op} is the operational temperature of the anaerobic digester (308.15 K), $p_{gas,tot}$ (bar) is the total gas pressure in the headspace, P_{atm} (bar) is the atmospheric pressure (1.013 bar) and Q_{gas} ($m^3 \cdot d^{-1}$) is the normalised gas flow rate (at P_{atm}) of produced gas. The number 16 represents the atomic weight of methane and t_{obs} is the length of the evaluation period (i.e., 364 days for BSM2).

6.3.7 Heating energy

The heating energy term (Technical Report No. 5) takes into account the energy demand HE ($kWh \cdot d^{-1}$) to heat the anaerobic digester and the heat generated from biogas electricity production. Assuming that 1 kg CH_4 produces 7 kWh of heat, the net heating energy demand can be calculated using:

$$HE^{net} = \max(0, HE - 7 \cdot MP)$$

The net heating energy cannot be negative so a correction term is added. It is assumed that the surplus heat that may be produced during electricity generation and that is not used for heating the anaerobic digester is not used elsewhere in the process.

The heating energy HE ($kWh \cdot d^{-1}$) needed to heat the flow of sludge fed to the anaerobic digester is calculated as the average energy input needed to heat the digester inlet sludge flow to the desired temperature in the anaerobic digester:

$$HE = \frac{24}{t_{obs} \cdot 86\,400} \int_{t_{start}}^{t_{end}} \rho_{H2O} \cdot c_{H2O} \cdot (T_{op} - T_{ad,i}(t)) \cdot Q_{ad}(t) \cdot \mathrm{d}t$$

where, ρ_{H2O} is the water density (1 000 $kg \cdot m^{-3}$), c_{H2O} is the specific heat capacity of water (4.186 $kJ \cdot kg^{-1} \cdot {}^\circ C^{-1}$), T_{op} is the operational temperature of the anaerobic digester (35°C or 308.15 K), $T_{ad,i}$ is the temperature of the anaerobic digester influent (expressed in the same units as T_{op}) and Q_{ad} is the flow rate to the anaerobic digester ($m^3 \cdot d^{-1}$). It is assumed that the sludge reaches the desired temperature within the digester hydraulic retention time and that heating power is constant. Heat losses to the surroundings *via* the digester walls are neglected in the calculation. Return liquors are assumed to loose the heat to the environment before being recycled to the water line at the influent and therefore do not heat the influent.

The temperature of the anaerobic digester influent is calculated as the flow weighted average of the temperature of the individual streams fed to the anaerobic digester, assuming the density and specific heat capacities to be constant and the same for all streams:

$$T_{ad,i}(t) = \frac{T_{pu}(t) \cdot Q_{pu}(t) + T_{tu}(t) \cdot Q_{tu}(t)}{Q_{ad}(t)}$$

where,

$$Q_{ad}(t) = Q_{pu}(t) + Q_{tu}(t)$$

The subscripts 'pu' and 'tu' refer to the primary clarifier underflow and thickener underflow, respectively.

6.4 CONTROLLER ASSESSMENT

In any given BSM implementation with a controller, the control behaviour must be assessed and reported. This assessment quantifies controller performance and can be divided into two categories:

- controlled variable performance;
- manipulated variable performance.

6.4.1 Controlled variable tracking

The controller's ability to track the set point is assessed by several measures based on e_j, the error in the controlled variable ($e_j(t) = Z_{j,setpoint}(t) - Z_{j,observed}(t)$) with the subscript j used to distinguish different controlled variables in the same system. The measures include: the integral of the absolute error, the integral of the squared error, the maximum deviation from the set point and the variance in the controlled variable error.

IAE (Integral of the Absolute Error):

$$IAE_i = \int_{t_{start}}^{t_{end}} \left| e_j(t) \right| dt$$

ISE (Integral of the Squared Error):

$$ISE_j = \int_{t_{start}}^{t_{end}} e_j(t)^2 \, dt$$

Maximum deviation from set point:

$$\max(Dev_j^{error}) = \max\left|e_j(t)\right|$$

Variance of the controlled variable error:

$$Var(e_j) = \overline{e_j^2} - \left(\overline{e_j}\right)^2$$

where,

$$\overline{e_j} = \frac{\int_{t_{start}}^{t_{end}} e_j(t)dt}{t_{obs}} \qquad \overline{e_j^2} = \frac{\int_{t_{start}}^{t_{end}} e_j(t)^2 dt}{t_{obs}}$$

6.4.2 Actuator performance

Although the controller's ability to track the set point is important, it is equally important to assess what impact the controller has on the actuator. This impact is assessed using several measures including the maximum deviation in the manipulated variable, the maximum deviation in the change in the manipulated variable and the variance in the change in the manipulated variable.

Maximum deviation in the manipulated variable:

$$\max(Dev_j^{MV}) = u_{j,max} - u_{j,min}$$

where, $u_j(t)$ is the value of the manipulated variable (MV) and the minimum and maximum are determined during the period of evaluation (i.e., 7 days for BSM1, 364 days for BSM2 and BSM1_LT). As above, the subscript j is used to distinguish different manipulated variables in the same system.

Maximum deviation in the change in the manipulated variable:

$$\max(Dev_j^{\Delta u_j}) = \max(\Delta u_j(t))$$

where,

$$\Delta u_j(t) = \left|u_j(t + \Delta t) - u_j(t)\right|, \text{ and } \Delta t = 0.25 \text{ h}$$

Variance in the change in the manipulated variable:

$$Var(\Delta u_j) = \overline{\Delta u_j^2} - \left(\overline{\Delta u_j}\right)^2$$

where,

$$\overline{\Delta u_j} = \frac{\int_{t_{start}}^{t_{end}} \Delta u_j(t)dt}{t_{obs}} \qquad \overline{\Delta u_j^2} = \frac{\int_{t_{start}}^{t_{end}} \Delta u_j(t)^2 dt}{t_{obs}}$$

6.4.3 Risk-related evaluation criteria

The risk assessment model provides new plant performance criteria related to the risk of microbiology-related settling problems in activated sludge systems. Although not as straightforward and easily validated as the effluent quality index and the operational cost index, the risk index can be used for relative comparisons of different control strategies as an indicator of potential settling problems that an otherwise excellent control strategy may lead to. The risk assessment is characterised by six different indices: risk of filamentous bulking due to low DO, risk of filamentous bulking due to low organic loading, risk of filamentous bulking due to nutrient deficiency, risk of foaming due to low F/M ratio, risk of foaming due to high readily biodegradable organic matter fraction and risk of rising sludge.

The six indices for the microbiology-related separation problems considered in the risk model can be integrated into a single overall risk index. To do that, an integrated value for filamentous bulking should be first obtained as the maximum value, at each time step, from the time series signals of the three risk indices for filamentous bulking problems (caused by low DO concentrations, low organic loading and nutrient deficiencies). At the same time, the integrated foaming index is determined as the maximum value, at each time step, of the foaming risk due to a low F/M ratio and the foaming risk due to a high readily biodegradable organic matter fraction. The final step simply consists of taking the maximum value at every time step from the integrated risk of bulking, the integrated risk of foaming and the risk of rising sludge to produce the overall risk index. For a specific activated sludge system, these integrated values give an indication of the overall risk of solids separation problems as well as an indication of which separation issue is the likely cause for concern.

The results from the risk assessment model are reported and quantified in four different ways for each of the settling problems and causes, for the integrated risks of bulking and foaming and for the overall risk index: (i) a time series plot (or data) showing the evolution of the risk occurrence for a specific settling problem (or for one of the integrated indices) during the evaluation period (0 indicates no risk and 1 means the highest possible risk); (ii) the percentage of time during which the model predicts risk for occurrence of a specific settling problem; (iii) the percentage of time during which the model predicts severe risk of settling problems (an arbitrary but customisable limit of ≥0.8 is used for defining a severe problem); and, (iv) the worst case scenario during the evaluation period, computed as the longest time interval the plant is exposed to an uninterrupted severe risk of experiencing a specific settling problem.

6.5 MONITORING PERFORMANCE ASSESSMENT

Aside from the evaluation of control strategies, the other work that was initiated with the BSM1_LT platform was the testing of process monitoring approaches. In order to evaluate and compare the usefulness or appropriateness of a fault detection approach, an objective way to evaluate the performance is needed. The design of this objective procedure has proven difficult due to the many monitoring methodologies available and due to the different objectives each of these methodologies pursues. For instance, it is difficult to use the same index to compare a neural network based classification algorithm for local sensor fault identification with a principal component analysis based method for global process fault detection. As a result, the development of a monitoring evaluation index for BSM1_LT has focused on finding a method to evaluate the performance of global approaches (i.e., comprising sensor, actuator and process faults in the whole system). The focus on global approaches is also reflected in the range of disturbances imposed on the system (Sections 4.3.3 and 4.4).

An objective index that evaluates the monitoring and diagnosis performance of different fault detection methods is proposed. This index penalises the fault detection method each time it fails. The fault detection method is evaluated at every sample point whether a fault in a sensor or actuator is occurring or not. The output of the fault detection method is the estimated state of the sensor or actuator. This output is compared to the true state to determine whether the output of the fault detection method is correct or not. When the output is not correct, the method is assigned penalty points. The more points a method gets, the worse the method has performed. The index is designed in such a way that:

- A penalty function is used at each time instant to penalise a false acceptance (FAC). The more time required to detect a fault event, the more penalty points will be assigned. Consequently, the proposed penalisation approach is dynamic and includes a means to evaluate the speed at which a fault detection method operates.
- The intermittent detection of a fault within the duration of a longer fault event is penalised further. The reason for this additional penalty is that it is highly undesirable for a detection method to switch between correctly and incorrectly detecting a fault as this can lead to a loss of trust of the method.
- A constant penalty value is assigned to each false alarm (FAL).

The instantaneous penalties are summed over time to obtain the cumulative penalties. It is assumed that slower event detection is worse than earlier detection and that loss of trust due to intermittent detection is problematic. The outcome of the evaluation index is a measure of reliability which ranges from 0% (not reliable) to 100% (reliable).

The details on how the fault detection index works are explained using the following figures. Figure 6.1 shows the evaluation of a hypothetical fault detection system to identify normal (state 1) and faulty (state 2) operation of a sensor over one day. As can be seen in this figure, the sensor fails from 0.5 to 0.8 days (Figure 6.1a, symbols). The hypothetical fault detection system provides an estimate of the sensor state (Figure 6.1a, dashed line) that is not correct all the time: the fault detection system indicates a false alarm (FAL) between 0.2 and 0.3 days and a false acceptance (FAC) between 0.5 and 0.6 days.

To evaluate the fault detection performance different steps are followed (Figure 6.1b–f):

Definition of a timer (k). An artificial timer (Figure 6.1b) is initialised at the beginning of a fault event (k(1) = 1). The timer is switched On (k(t) = k(t – 1) + 1) when a fault event starts (0.5 days in this example) and is switched Off (k(t) = 1) at the end of the event as can be seen in Figure 6.1b.

Recall that methods that intermittently detect a fault are further penalised (Figure 6.2). To do this, the time from the start of the fault, k(t) is artificially augmented with a positive increment k_{switch} whenever the alarm switches from correct detection to false acceptance (i.e., k(t) becomes k(t) + k_{switch}).

Calculation of the penalty function (P(t)). Figure 6.1c shows the penalty function for false acceptance (P_{FAC}), which is calculated using the artificial timer values from Figure 6.1b and:

$$P_{FAC}(t) = P_{FAC,0} + (P_{FAC,sat} - P_{FAC,0}) * \left(1 - e^{\left(\frac{-k(t)}{\tau_{FAC}}\right)}\right)$$

The penalty function includes a saturation function and reaches the maximal penalty level ($P_{FAC,sat}$) after a given time (τ_{FAC}). τ_{FAC} is set to reflect the urgency of detection (in this example τ_{FAC} = 45 min.). The penalty function for false alarms (P_{FAL}) is a constant (Figure 6.1d).

$$P_{FAL}(t) = P_{FAL,0}$$

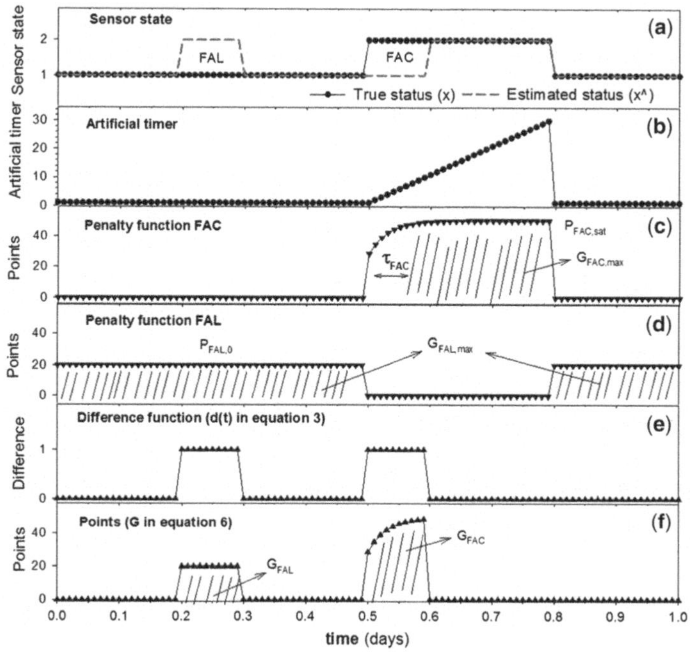

Figure 6.1 Representation of the fault detection index concept.

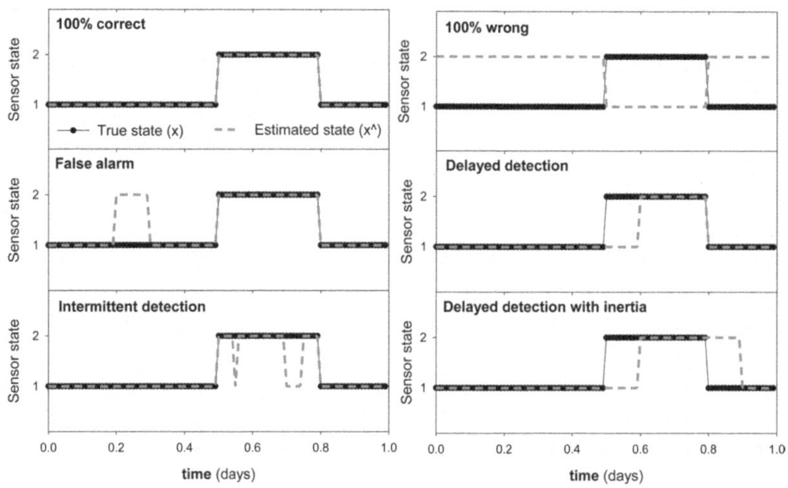

Figure 6.2 System (symbols) and diagnosed sensor status (dashed line) for five artificial methods (state 1 signifies a correctly operating sensor and state 2 is a faulty sensor).

Calculation of penalty points (G). The penalties assigned to the evaluated fault detection system are obtained as follows. First, the difference between the true (x) state and the estimated ($\hat{x}$) state (Figure 6.1e) is computed (0 when correct, 1 if incorrect). Then, this sequence of zeroes and ones is multiplied by the penalty functions ($P_{FAC}(t)$ and $P_{FAL}(t)$). The areas under the function shown in Figure 6.1f are the accumulated penalties (G_{FAC} and G_{FAL}).

$$d(t) = \max[0, abs(x(t) - \hat{x}(t))]$$

$$G_{FAC} = \sum_{t_0}^{t_{end}} [P_{FAC}(t)*d(t)]$$

$$G_{FAL} = \sum_{t_0}^{t_{end}} [P_{FAL}(t)*d(t)]$$

$$G = G_{FAC}(t) + G_{FAL}(t)$$

Calculation of maximum penalty. The maximum penalty (G) and the maximum penalties for FAC and FAL ($G_{FAC,max}$ and $G_{FAL,max}$, respectively; Figure 6.1c,d) are obtained by setting $d(t)$ to 1 for all time instants. The sum of P_{FAC} and P_{FAL} penalty functions represents the reference case that results from maximum penalisation at all evaluation times.

$$G_{FAC,max} = \sum_{t_0}^{t_{end}} [P_{FAC}(t)]$$

$$G_{FAL,max} = \sum_{t_0}^{t_{end}} [P_{FAL}(t)]$$

$$G_{max} = G_{FAC,max} + G_{FAL,max}$$

Measure of reliability (J). J represents how reliable the monitoring system is at detecting faults. J is calculated using:

$$J = \left(1 - \frac{G_{FAC} + G_{FAL}}{G_{FAC,max} + G_{FAL,max}}\right) \times 100$$

$$J_{PAC} = \left(1 - \frac{G_{FAC}}{G_{FAC,max}}\right) \times 100$$

$$J_{FAL} = \left(1 - \frac{G_{FAL}}{G_{FAL,max}}\right) \times 100$$

The fault detection system is assumed to be more reliable the closer J is to 100. More information can be extracted from J_{FAC} and J_{FAL}, which indicate the reliability of the detection system with respect to the generation of false acceptance and false alarms, independently.

In summary, the inputs to the fault detection evaluation index are the true and estimated sensor/actuator states provided by the fault detection system that is being evaluated. The outputs of the monitoring performance evaluation system are expressed as measures of reliability of the fault detection system (J). Five parameters need to be specified for the index: $P_{FAC,0}$, $P_{FAC,sat}$, τ_{FAC}, $P_{FAL,0}$ and k_{switch}. The base level of penalisation, which is time-independent, is indicated by $P_{FAC,0}$ and $P_{FAL,0}$. $P_{FAC,sat}$ is used to indicate the maximal penalty level and τ_{FAC} reflects the urgency of penalising the method for the late detection of a fault (Figure 6.1c, $\tau_{FAC} = 45$ min.).

As evaluation experience with this index is still being developed, further tuning of the performance evaluation parameters is needed. However, it should be kept in mind that the selection of these parameters of the index should reflect the benefits and costs associated with the implementation of a fault detection system. Although in the given example only two types of sensor states are considered (normal and faulty) it is possible to extend this index to different types of faults, making this index approach consistent with the benchmark philosophy of providing a platform for investigating new and innovative approaches to simulation problems.

6.6 EVALUATION SUMMARY

An attempt has been made to develop evaluation criteria that are applicable to all of the benchmark models while at the same time providing insight into the evaluation of an implemented control strategy or monitoring methodology. Irrespective of what is being evaluated, the criteria provide a geographically independent means to compare the behaviour of different strategies. Table 6.2 outlines the criteria that are to be calculated for each benchmark model.

Table 6.2 BSM evaluation criteria to be calculated for each model.

Criterion	BSM1	BSM1_LT	BSM2
IQI	✓	✓	✓
EQI	✓	✓	✓
Effluent			
95 percentile	✓	✓	✓
number of violations	✓	✓	✓
violation time	✓	✓	✓
OCI			
aeration energy	✓	✓	✓
pumping energy	✓	✓	✓
sludge production	✓	✓	✓
external carbon	✓	✓	✓
mixing energy	✓	✓	✓
methane production			✓
heating energy			✓
Controller assessment	✓	✓	✓
Risk assessment	✓	✓	✓
Fault detection index		✓	

Although the evaluation criteria are meant to be geographically independent, the structure of the performance measures allows for location specific criteria to be defined in subsequent analyses. That is, the performance measures described above *must* be calculated for each strategy simulation, but emphasis can be placed on specific performance terms depending on location specific criteria if a user so wishes. For example, for a particular user if effluent quality is of primary concern irrespective of overall costs then the analysis of the performance index terms can be weighted accordingly. Alternatively, in another location where reducing overall costs is the primary objective, the index can be tailored to that situation. This structure allows for substantial flexibility in applying the *simulation benchmark* to specific control and monitoring strategies, while at the same time providing a means to make meaningful location specific comparisons and design decisions.

Chapter 7

Simulation procedure

J. B. Copp, K. V. Gernaey, U. Jeppsson and P. A. Vanrolleghem

The descriptions of the BSM simulation procedures are rigid to ensure the consistent application of the benchmark and to ensure that similar analyses are done on the output data. It is therefore critical that exactly the same simulation procedure be used for each simulated control or monitoring strategy evaluation so that an objective comparison can be made.

7.1 BSM1

For BSM1, a two-step simulation procedure is defined that involves simulations to steady state followed by dynamic simulations using the three influent data files described in Chapter 4.

Steady state simulations

The initial step in the simulation procedure is to simulate the system under study to steady state using an influent of constant flow and composition. The flow-weighted *dry weather* data of Table 7.1 are used for this purpose and steady state is defined using either the software steady state solver *or* by simulating at least 100 days using a constant influent (possibly taking advantage of certain variable step numerical solvers to speed up the calculations). All dynamic simulations should follow a steady state simulation. The steady state simulation should include the same control strategy the user wants to benchmark. However, inactivation of all noise and use of ideal sensor and actuator models should be considered as this will speed up the steady state convergence and will not influence the steady state results. This steady state procedure ensures a consistent starting point and should eliminate the influence of initial conditions on the generated dynamic output. If the investigated control strategy is based on non-stationary control actions (e.g., intermittent aeration) then the non-steady operation should be used during this period to achieve a 'pseudo' steady state before initiating the dynamic simulations.

Dynamic simulations

Next, dynamic simulations should be performed using the influent files described previously. The implementation of these dynamic simulations will vary with the simulator being used, however a general overview of the required procedure is outlined in this section.

Table 7.1 Steady state influent for BSM1.

Variable	Value	Unit
$Q_{i,stab}$	18 446	$m^3 \cdot d^{-1}$
$Q_{w,stab}$	385	$m^3 \cdot d^{-1}$
$S_{I,stab}$	30	$g\ COD \cdot m^{-3}$
$S_{S,stab}$	69.5	$g\ COD \cdot m^{-3}$
$X_{I,stab}$	51.2	$g\ COD \cdot m^{-3}$
$X_{S,stab}$	202.32	$g\ COD \cdot m^{-3}$
$X_{B,H,stab}$	28.17	$g\ COD \cdot m^{-3}$
$X_{B,A,stab}$	0	$g\ COD \cdot m^{-3}$
$X_{P,stab}$	0	$g\ COD \cdot m^{-3}$
$S_{O,stab}$	0	$g\ O_2 \cdot m^{-3}$
$S_{NO,stab}$	0	$g\ N \cdot m^{-3}$
$S_{NH,stab}$	31.56	$g\ N \cdot m^{-3}$
$S_{ND,stab}$	6.95	$g\ N \cdot m^{-3}$
$X_{ND,stab}$	10.59	$g\ N \cdot m^{-3}$
$S_{ALK,stab}$	7.00	$mol\ HCO_3 \cdot m^{-3}$
T_{stab}	15.00	°C

Starting from the steady (or 'pseudo' steady) state solution, using the *dry weather* influent file as a dynamic input, the system under study should be simulated for 14 days (including noise and non-ideal sensor and actuator models). The resulting state variable values should then be saved (if possible, in the simulator being used) for all unit processes. These state variable values represent the starting point for evaluating the dynamic response of the plant to each of the three influent disturbance files. From the state achieved above, simulate a further 14 days using the *dry weather*, *storm event* and *rain event* influent files in separate simulation studies, but each time starting from the state achieved after the initial 14-day *dry weather* simulation. That is, for any one system at steady state, there are three 28-day dynamic simulations to perform: *dry-dry*, *dry-storm* and *dry-rain*. Where feasible, the user can save all conditions at the end of the first 14-day *dry weather* simulation, reinitiate those conditions in each case and run three 14-day simulations for analysis (*dry, storm, rain*).

The output data generated from the simulations described above are used to examine the dynamic performance of the process. The data of interest from these dynamic simulations are the data generated during the last 7 days of dynamic simulation. That is, in this scenario with 28-day simulations, the data of interest are from the beginning of day 22 until the end of day 28 and include three data sets: one for the *dry weather* simulation, one for the *storm event* simulation and one for the *rain event* simulation. Output data should be recorded at 15-minute intervals for each variable of interest. If the simulation time steps are not exactly aligned with the 15-minute intervals, linear interpolation between simulation time steps should be adopted to calculate the simulation output values at each 15 minute interval exactly.

7.2 BSM1_LT

BSM1_LT was designed as a tool for control strategy or monitoring algorithm evaluation using an evaluation period of one year. The simulation protocol for BSM1_LT is as follows: First, the model is allowed to reach steady state *via* the software steady state solver or by running a 200-day simulation with a constant influent. In either case, the simulations should be run without any faults or inhibition. The simulation should be based on the selected control strategy but noise can be deactivated and ideal sensor and actuator models should be used. As the steady state simulation is based on a constant input (Table 7.2) approximating average conditions (using a temperature equal to the BSM1 temperature, 15°C) any time-based control related to temperature effects should be disabled (e.g., seasonal control of the wastage flow rate) and purposefully set to the values the active controllers will have when starting the dynamic simulation. Afterwards, a dynamic simulation is conducted for a period of 609 days using the defined dynamic influent data (flows, concentrations and temperature as defined in Chapter 4). The dynamic simulation is run for 245 days: the first 63 days allows the system to reach a dynamic 'pseudo' steady-state, the next 182 days creates data for training the control or process monitoring algorithms. The first 245 days can be run as many times as desired to tune the algorithms. The final 364 days is used for control and monitoring performance evaluation (days 245–609).

Table 7.2 Influent values for the 200-day stabilisation period when using BSM1_LT.

Variable	Value	Unit
$Q_{i,stab}$	20 850.54	$m^3 \cdot d^{-1}$
$Q_{w,stab}$	300	$m^3 \cdot d^{-1}$
$S_{I,stab}$	27.90	$g\ COD \cdot m^{-3}$
$S_{S,stab}$	54.68	$g\ COD \cdot m^{-3}$
$X_{I,stab}$	48.29	$g\ COD \cdot m^{-3}$
$X_{S,stab}$	190.20	$g\ COD \cdot m^{-3}$
$X_{B,H,stab}$	26.48	$g\ COD \cdot m^{-3}$
$X_{B,A,stab}$	0	$g\ COD \cdot m^{-3}$
$X_{P,stab}$	0	$g\ COD \cdot m^{-3}$
$S_{O,stab}$	0	$g\ O_2 \cdot m^{-3}$
$S_{NO,stab}$	0	$g\ N \cdot m^{-3}$
$S_{NH,stab}$	22.88	$g\ N \cdot m^{-3}$
$S_{ND,stab}$	5.43	$g\ N \cdot m^{-3}$
$X_{ND,stab}$	8.00	$g\ N \cdot m^{-3}$
$S_{ALK,stab}$	7.00	$mol\ HCO_3 \cdot m^{-3}$
T_{stab}	15.00	°C
$S_{tox,stab}$	0	$g \cdot m^{-3}$
$X_{tox,stab}$	0	$g \cdot m^{-3}$

Default control strategies for the reference case are: (1) NO_3-N concentration (class B_0 sensor) maintained in the second compartment at a predetermined set point value (1 g NO_3-N $\cdot m^{-3}$) by manipulating the internal recycle rate; and, (2) dissolved oxygen concentration (class A sensor) control in the fifth compartment at a predetermined set point value (2 g $O_2 \cdot m^{-3}$) by manipulation of K_La (constrained between 0 and 360 d^{-1}). In addition to these two controllers, carbon, with a concentration of 400 000 g COD $\cdot m^{-3}$, is added to the first reactor at a fixed rate of 2.0 $m^3 \cdot d^{-1}$. The only other difference with these strategies as compared to the BSM1 definition is the control of the wastage rate. In BSM1, the wastage rate is constant (due to the short evaluation time), in BSM1_LT, two different flow rates are imposed depending on the time of the year (i.e., warm and cold seasons, see Table 7.3).

Table 7.3 BSM1_LT wastage flow rate as a function of simulation time.

Time (d)	Q_w ($m^3 \cdot d^{-1}$)
$0 \le t < 182$	300
$182 \le t < 364$	400
$364 \le t < 546$	300
$546 \le t \le 609$	400

When the evaluation of process monitoring methods is the goal, sensor and actuator fault models are activated together with inhibitory and toxic material in the influent wastewater. In this case, the predefined input file now contains sensor and actuator fault sequences (obtained from the Markov chain models, Chapter 4) and toxicity events.

7.3 BSM2

The simulation procedure for BSM2 is straightforward. The system is first simulated for at least 200 days using the predefined constant influent data to reach a steady state. The constant influent represents flow-weighted average influent values corresponding to the one year evaluation period; namely, the last 364 days of the 609-day dynamic input data of the BSM2 (Table 7.4). The simulation should be based on the selected control strategy, but noise can be deactivated and ideal sensor and actuator models used. As with BSM_LT, because the steady state simulation is based on a constant input approximating average conditions (temperature equal to 15°C), any time-based control related to temperature effects should be disabled (e.g., season-based control of the wastage flow rate) and purposefully set to the values the active controllers will have when starting the dynamic simulation. If a non-stationary control strategy is used then only a 'pseudo' steady state can be achieved. This procedure ensures a consistent starting point and should eliminate the influence of initial conditions on the generated dynamic output.

Table 7.4 Influent values for the 200 day stabilisation period for BSM2.

Variable	Value	Unit
$Q_{i,stab}$	20 648.36	$m^3 \cdot d^{-1}$
$Q_{w,stab}$	300	$m^3 \cdot d^{-1}$
$S_{I,stab}$	27.23	$g\ COD \cdot m^{-3}$
$S_{S,stab}$	58.18	$g\ COD \cdot m^{-3}$
$X_{I,stab}$	92.50	$g\ COD \cdot m^{-3}$
$X_{S,stab}$	363.94	$g\ COD \cdot m^{-3}$
$X_{B,H,stab}$	50.68	$g\ COD \cdot m^{-3}$
$X_{B,A,stab}$	0	$g\ COD \cdot m^{-3}$
$X_{P,stab}$	0	$g\ COD \cdot m^{-3}$
$S_{O,stab}$	0	$g\ O_2 \cdot m^{-3}$
$S_{NO,stab}$	0	$g\ N \cdot m^{-3}$
$S_{NH,stab}$	23.86	$g\ N \cdot m^{-3}$

(Continued)

Table 7.4 Influent values for the 200 day stabilisation period for BSM2 (*Continued*).

Variable	Value	Unit
$S_{ND,stab}$	5.65	$g\,N \cdot m^{-3}$
$X_{ND,stab}$	16.13	$g\,N \cdot m^{-3}$
$S_{ALK,stab}$	7.00	$mol\,HCO_3 \cdot m^{-3}$
T_{stab}	14.86	°C

The steady state values obtained in this first simulation are subsequently used as initial values for simulations using the dynamic influent and the wastage flow rates of Table 7.5. From this starting point, BSM2 is simulated for 63 days using the dynamic influent with controls active (including noise and non-ideal sensor and actuator models) to achieve a 'pseudo' steady state. This period is followed by 182 days of dynamic simulation. This 'training' period is used to tune adaptive or model-based controllers, to set internal parameters or to tune the control algorithms in another way. Once training is complete, BSM2 is simulated for an additional 364 days with the output data (stored at 15-minute intervals) used for plant performance evaluation. This final period is only simulated once the algorithms are suitably tuned/trained and need not be simulated while testing. Each control strategy, regardless of the study objective, is simulated in precisely the same way to ensure standardised results.

Table 7.5 BSM2 wastage flow rate as a function of simulation time.

Time (d)	Q_w ($m^3 \cdot d^{-1}$)
$0 \le t < 182$	300
$182 \le t < 364$	450
$364 \le t < 546$	300
$546 \le t \le 609$	450

Chapter 8

Ring-testing

I. Nopens, W. De Keyser, L. Corominas, L. Benedetti, M.-N. Pons, J. Alex, J. B. Copp, J. Dudley, C. Rosen, P. A. Vanrolleghem and U. Jeppsson

From the outset, the primary objective of the BSM development has been to create a tool that could be used to compare different wastewater treatment control strategies. Many control strategies have been proposed in the literature; however, the literature does not provide a clear basis for comparison of these strategies because of the many confounding influences that have an impact on the system. Many of these influences are easily recognised. For instance, physical characteristics of the process can have an impact on process performance, which makes the comparison of strategies applied to different reactor layouts difficult. However, there are more subtle differences as well. For instance, the simulated process impact will be a function of the model(s) used. It will be a function of the parameters and it will be a function of the model implementation. Even though a given literature study indicates that a particular model was used, experience tells us that commercially available simulation software packages have specific features that can impact the model output.

The BSM concept is built on the premise that all users generate output that is comparable regardless of the simulation software they are using. To do this, it is critical that all studies begin from a verified and identical starting point. This may seem trivial as the models and parameters are seemingly all defined within this context. However, it is not. Platform-specific and model implementation issues result in different output even when the models and parameters are identical. This chapter addresses the issue of cross-platform testing and the substantial amount of effort that has gone into verifying (and correcting, where errors were found) the output from different simulation platforms. So far, some, and in some cases all, of the various BSM results have been achievable using BioWin, EFOR, GPS-X, Matlab/Simulink, Simba, STOAT, WEST and a user defined FORTRAN code (Alex *et al.* 1999; Pons *et al.* 1999).

Although this process of cross-platform testing to achieve substantively identical results has been a time consuming exercise, it has provided a means to develop a significant insight into the simulators and the simulation process. Furthermore, as each of the simulators requires a different method of implementation, each of the simulators has required simulator-specific alterations and fine-tuning to get agreement in the output data. Knowledge of these simulator-specific alterations is crucial for benchmark use and is discussed in detail elsewhere (Copp, 2002). It should be made clear that cross-platform information and testing is not meant to be used or interpreted as a comparative study of different simulators.

During cross-platform testing, each simulator was subjected to the same tests and the results were compared. These comprehensive tests included the comparison of individual sub-model results (i.e., primary clarifier, ADM1, …) as well as the BSMs as a whole. In each case, the BSM (or sub-model) was implemented by different people and the simulation conditions were fully defined. Both steady state and dynamic conditions were tested and when the results varied, the reasons for those differences were investigated. This effort identified errors in the model code in almost all cases (these errors were passed on to the respective software developers and have since been corrected) and also highlighted the impact of differences in simulator-specific issues like integration routines and the way influent components are interpolated. Some of these differences resulted in substantial deviations in output, but other minor variances in the output were also identified. It is important to recognise that although minor differences may not have an impact on the conclusions from a practical engineering study, they are important for this work. Duplicating these steady state and dynamic results is an essential first step in the evaluation procedure and ensures that the simulator being used is tuned according to the BSM specifications, which in turn should ensure the consistent comparison of control strategy results.

8.1 STEADY STATE VERIFICATION

For any given model (whether that be the whole BSM2, or a simple sub-model), the first step is to implement the model in the simulator of choice. The constant influent to the model is defined according to the BSM description and the model is simulated to steady state using either the software steady state solver or simulating for 100 days (BSM1) or at least 200 days (BSM1_LT and BSM2) with all model components held constant (open loop, i.e., no active control).

The generated output is then compared to the verified output and the model is assumed correct if the model results match, or in error if the results do not match. Some BSM1 steady state output results are listed in Table 8.1 (more results can be found in the appendices and the complete set of results can be found in Technical Report no. 14). These results have been duplicated using BioWin, EFOR, GPS-X, Matlab/Simulink, Simba, STOAT, WEST and a user defined FORTRAN code and for that reason are assumed to be correct. Ring-testing of BSM1_LT is on-going in Matlab/Simulink, WEST and values will be available in an updated technical report. For BSM2 the ring-testing was successfully performed using Matlab/Simulink, Simba, WEST and a user defined FORTRAN code.

It is assumed that the simulator and associated models being used have been input correctly once similar steady state results have been attained. The results obtained with the simulators listed above were the same to the 4th or 5th significant digits for all states in all streams indicating that essentially identical results can be achieved when the simulators are set up properly. The details of these results can be found in the technical report.

If, during implementation, the achieved results do not fall within those tolerances, users are advised to re-examine their set-up looking for possible errors. Note that tuning experience has shown that these discrepancies may be the result of user-input errors (i.e., incorrect parameters, incorrectly specified flow rates, …) or simulator-specific options (i.e., convergence tolerances of the integrator, …). Users may need to examine both possibilities to find a particular problem.

Once acceptable steady state values have been achieved, users are encouraged to perform the dynamic simulations to further test the tuning of their simulator.

BSM2 consists of a number of sub-models (primary clarifier, thickener, ASM to ADM interface, anaerobic digester, ADM to ASM interface, dewatering unit), and as such, it is recommended that users verify individual sub-model implementations prior to the entire BSM2 implementation. Verification of a sub-model implementation is relatively straightforward compared to verification of the full BSM2. This approach provides a means to detect and correct errors in the sub-models without having to be concerned with the impact that an error in one process might have on another. This approach is valid irrespective of the BSM being used. After verification of the sub-models, the entire BSM system of interest can be connected together and the output checked. Trying to debug a BSM implementation without first verifying individual sub-models is not recommended and will be time consuming.

Table 8.1 Steady state simulation results for BSM1 in open loop based on average dry weather data.

Variable	Unit	Tank 1	Tank 5	Settler underflow	Effluent
Q_e	$m^3 \cdot d^{-1}$				18 061
S_I	$g\ COD \cdot m^{-3}$	30.00	30.00	30.00	30.00
S_S	$g\ COD \cdot m^{-3}$	2.807	0.889	0.889	0.889
X_I	$g\ COD \cdot m^{-3}$	1 149	1 149	2 247	4.392
X_S	$g\ COD \cdot m^{-3}$	82.14	49.31	96.42	0.188
$X_{B,H}$	$g\ COD \cdot m^{-3}$	2 552	2 559	5 005	9.782
$X_{B,A}$	$g\ COD \cdot m^{-3}$	148.4	149.8	292.9	0.573
X_P	$g\ COD \cdot m^{-3}$	448.9	452.2	884.3	1.728
S_O	$g\ O_2 \cdot m^{-3}$	0.004	0.491	0.491	0.491
S_{NO}	$g\ N \cdot m^{-3}$	5.369	10.41	10.41	10.41
S_{NH}	$g\ N \cdot m^{-3}$	7.919	1.735	1.735	1.735
S_{ND}	$g\ N \cdot m^{-3}$	1.217	0.689	0.689	0.689
X_{ND}	$g\ N \cdot m^{-3}$	5.285	3.527	6.897	0.013
S_{ALK}	$mol\ HCO_3 \cdot m^{-3}$	4.928	4.126	4.126	4.126
TSS	$g\ SS \cdot m^{-3}$	3 285	3 270	6 394	12.50
TKN	$g\ N \cdot m^{-3}$	326.3	318.7	621.0	3.630
N_{tot}	$g\ N \cdot m^{-3}$	331.7	329.1	631.4	14.04
COD_{tot}	$g\ COD \cdot m^{-3}$	4 413	4 390	8 556	47.55
BOD_5	$g\ O_2 \cdot m^{-3}$	642.3	635.6	1 242	2.651
SRT	d		for the complete BSM1: 9.18		
HRT	h		for the complete BSM1: 15.61		

8.2 DYNAMIC VERIFICATION

To test the dynamic implementation, it may be necessary to perform a series of simulations using the procedures described previously. In the BSM1 case, the three dynamic influent files are used in separate simulations with the uncontrolled plant (open loop). In the BSM2 case, the open-loop scenario is simulated for 609 days using the BSM2-defined influent. Once completed, the various evaluation criteria should be calculated and then compared to the accepted evaluation quantities defined in the BSM descriptions. The recognised performance index results for BSM1 and BSM2 are listed in Tables 8.2 and 8.3 and a complete listing of the compiled dynamic results is included in Technical Report No. 14. Once acceptable dynamic results are achieved, the user can be reassured that the simulator being used is tuned in accordance with the benchmark specifications.

Table 8.2 Effluent average concentrations for BSM1 (dry weather conditions) and BSM2, both in open loop (left). Evaluation criteria for BSM1 (dry weather conditions) and BSM2, both in open loop (right). Note that the numbers presented here are the average results from 5 simulators (BSM1) and 4 simulators (BSM2), respectively.

Variable	Unit	BSM1	BSM2	Variable	Unit	BSM1	BSM2
Q_e	$m^3 \cdot d^{-1}$	18 060	20 660	IQI	$kg\ PU \cdot d^{-1}$	52 085	74 790
S_I	$g\ COD \cdot m^{-3}$	30.00	28.05	EQI	$kg\ PU \cdot d^{-1}$	6 694	5 659
S_S	$g\ COD \cdot m^{-3}$	0.972	0.811	OCI	–	16 152	9 221
X_I	$g\ COD \cdot m^{-3}$	4.580	6.490	*AE*	$kWh \cdot d^{-1}$	3 341	4 000
X_S	$g\ COD \cdot m^{-3}$	0.223	0.348	*PE*	$kWh \cdot d^{-1}$	388.2	441.4
$X_{B,H}$	$g\ COD \cdot m^{-3}$	10.22	9.692	*SP*	$kg\ SS \cdot d^{-1}$	2 462	2 654
$X_{B,A}$	$g\ COD \cdot m^{-3}$	0.542	0.675	*EC*	$kg\ COD \cdot d^{-1}$	0.0	800.0
X_P	$g\ COD \cdot m^{-3}$	1.758	3.991	*ME*	$kWh \cdot d^{-1}$	240.0	768.0
S_O	$g\ O_2 \cdot m^{-3}$	0.758	1.032	*HE*	$kWh \cdot d^{-1}$	–	4 177
S_{NO}	$g\ N \cdot m^{-3}$	8.829	7.473	*MP*	$kg\ CH_4 \cdot d^{-1}$	–	1 059
S_{NH}	$g\ N \cdot m^{-3}$	4.769	1.648	$S_{NH,e95}$	$g\ N \cdot m^{-3}$	8.640	4.642
S_{ND}	$g\ N \cdot m^{-3}$	0.728	0.602	$N_{tot,e95}$	$g\ N \cdot m^{-3}$	18.34	15.11
X_{ND}	$g\ N \cdot m^{-3}$	0.016	0.020	TSS_{e95}	$g\ SS \cdot m^{-3}$	15.73	20.13
S_{ALK}	$mol\ HCO_3 \cdot m^{-3}$	4.460	4.796	***Violations****			
TSS	$g\ SS \cdot m^{-3}$	12.99	15.90	N_{tot} viol.	d	0.463	0.375
TKN	$g\ N \cdot m^{-3}$	6.754	3.728		%	6.616	0.103
N_{tot}	$g\ N \cdot m^{-3}$	15.58	11,20		–	5	4
COD_{tot}	$g\ COD \cdot m^{-3}$	48.29	50.06	COD_{tot} viol.	d	0.0	0.219
BOD_5	$g\ O_2 \cdot m^{-3}$	2.774	2.766		%	0.0	0.060
					–	0	3
				S_{NH} viol.	d	4.306	29.95
					%	61.50	8.229
					–	7	170
				TSS viol.	d	0.0	1.435
					%	0.0	0.394
					–	0	12
				BOD_5 viol.	d	0.0	0.812
					%	0.0	0.223
					–	0	8

**limits are:* $S_{NH} = 4\ g\ N \cdot m^{-3}$; $TSS = 30\ g \cdot m^{-3}$; $N_{tot} = 18\ g\ N \cdot m^{-3}$; $COD_{tot} = 100\ g\ COD \cdot m^{-3}$; $BOD_5 = 10\ g\ O_2 \cdot m^{-3}$; violations are expressed in days, % of total evaluation period, median number of limit violations. $HE^{net} = 0$ for BSM2 as more heat energy is produced by the system than needed by the digester and PU is pollution units.

Table 8.3 Settling risk results for BSM1 (dry weather conditions) and BSM2, both in open loop.

Individual risk	BSM1	BSM2
Bulking due to low DO values (average risk)	0.323	0.359
Bulking due to low DO values (%time in severe risk, >0.8)	8.185	0.000
Bulking due to low organic loading (average risk)	0.746	0.890
Bulking due to low organic loading (%time in severe risk, >0.8)	42.41	73.61
Bulking due to nutrient deficiency (average risk)	0.0001	0.0001
Bulking due to nutrient deficiency (%time in severe risk, >0.8)	0.000	0.000
Foaming due to low F/M ratio (average risk)	0.543	0.890
Foaming due to low F/M ratio (%time in severe risk, 0.8)	0.000	73.60
Foaming due to high RBOM* fraction (average risk)	0.026	0.003
Foaming due to high RBOM* fraction (%time in severe risk, >0.8)	0.000	0.000
Rising sludge (average risk)	0.682	0.436
Rising sludge (%time in severe risk, >0.8)	34.08	11.87
Integrated risks		
Bulking (average risk)	0.810	0.890
Bulking (%time in severe risk, >0.8)	50.60	73.61
Foaming (average risk)	0.550	0.890
Foaming (%time in severe risk, >0.8)	0.000	73.60
Rising (average risk)	0.682	0.436
Rising (%time in severe risk, >0.8)	34.08	11.87
Overall Risk (average risk)	0.888	0.892
Overall Risk (%time in severe risk, >0.8)	65.33	74.28

**Readily biodegradable organic matter*

Readers are reminded that the OCI includes a net heating energy term, HE^{net} (Chapter 6). The HE term in Table 8.2 refers to the energy required to heat the anaerobic digester influent, but HE^{net} (max (0, $HE - 7 \cdot MP$)) includes a credit for heat generated from the produced biogas (only used for heating the AD system). Therefore, HE^{net} should not be included when calculating the OCI from the values in this table.

A substantial amount of work has gone into evaluating the dynamic responses of various software packages and it has been determined that it is not realistic for each simulator to produce precisely the same instantaneous dynamic results (unlike the steady state condition which should be reproducible using all simulators). Not all of the simulators handle dynamic events in the same way and the numerical engine in each simulator is somewhat different. Furthermore, it is not always feasible to alter the specific features in commercially available simulators. Some consideration has to be given to the possibility that the instantaneous dynamic data will be different. For instance, some simulators allow the user to specify specific communication intervals (i.e., every 15 minutes) and the simulator will calculate output at precisely that time stamp. Other simulators do not necessarily calculate the output at specific times, but rather interpolate the output from points on either side of the required time stamp. Issues like this make the comparison of instantaneous dynamic results meaningless. However, because these differences are anticipated to average out if the simulation time is long enough, average concentrations and the evaluation criteria offer a suitable substitute to check the dynamic response. Using the evaluation criteria terms as a measure of the simulated dynamic behaviour, it is possible to determine whether the particular simulator being used is dynamically synchronised with the output of the many simulators that have verified the available dynamic performance data.

8.3 FINDINGS

The BSM1 ring-test results were excellent for steady state as well as dynamic conditions (dry, rain and storm weather) and proved to be an extremely valuable exercise for debugging the commercial simulators. This also indicates that all the current implementations are essentially identical. The BSM2 ring-testing process was performed in several steps, but again, excellent agreement was achieved. The activated sludge portion of BSM2 (AS tanks and secondary clarifier) is identical to BSM1 (apart from reactor volumes and temperature dependency) and was ring-tested in that exercise. All the other BSM2 unit processes (primary clarifier, thickener, ASM1 to ADM1 interface, ADM1, ADM1 to ASM1 interface and dewatering unit) were ring-tested separately under steady state conditions. The BSM2 influent generator (Gernaey *et al.* 2011) was tested using two independent Matlab/Simulink implementations and these independent tests produced identical results. Ring-testing of the sensor models with noise produced equally successful results as described in the detailed ring-testing technical report. The Matlab/Simulink and Simba implementations gave essentially identical results up to machine precision, but as they use the same numerical engine this was not surprising. The WEST and Fortran simulation results differed slightly (relative difference of 10^{-5}), but were deemed very close given the challenges presented when simulating noise. Once the implementation of each unit process was verified, and deemed properly implemented, the sub-models were connected and the results from BSM2 as a whole were compared. In general, the obtained results were very good with only minor differences. Work on ring-testing BSM1_LT is ongoing. Portions of that implementation are done (i.e., control) and the results have been finalised, while others (i.e., fault detection and toxicity) are still in progress.

The ring-testing work highlighted several issues. For instance, the more complex the models, the larger the deviations tended to get. The deviations were somewhat larger in BSM2 when compared to the BSM1 results and the deviations, for the most part, seemed to originate in the sludge train. The ASM1 to ADM1 interface for instance is sensitive to numerical errors as the absolute values of some of the calculated quantities are small (e.g., inorganic carbon concentration). These deviations propagate to the anaerobic digester resulting in small differences in its gaseous output. Although this did not have significant repercussions on the general behaviour of BSM2, it did impact the evaluation criteria. Other deviations were attributed to numerical methods rather than to differences in model implementations. And, under dynamic conditions these small deviations did not result in substantial differences in the average system behaviour. The computed evaluation criteria tended to be more sensitive to these small deviations because of the longer evaluation period. An analysis of the errors indicated that the longer evaluation period in BSM1_LT and BSM2 caused the accumulation of errors which resulted in noticeable deviations in the performance criteria (based on $364*24*4 = 34\,944$ measurement intervals). Tests to steady state, in open loop, and in closed loop with both ideal and realistic sensors all showed very similar outcomes leading to the conclusion that the model implementations were the same in each platform tested and confirming that with the same implementation, similar results can be achieved in different simulators.

Although every attempt has been made to generate identical results, it was shown that it is not possible to get absolutely identical results under dynamic conditions, especially in closed loop as the controllers were implemented and tuned individually for each platform without any Task Group specification.

As a result, it may be necessary to consider the simulator as a variable for any cross-platform comparison of control strategy results. To avoid this problem, it is recommended that the benchmarking of control strategies by any one user be done in a single platform. It is assumed that simulator-specific numerical issues will be similar in different simulations, resulting in their effect being cancelled out when comparing performance of different control strategies. Note that the platform effect should be considered small and would only impact conclusions regarding control strategies that have very similar performance.

Chapter 9

BSM limitations

J. B. Copp, K. V. Gernaey, U. Jeppsson and P. A. Vanrolleghem

The use of the benchmark systems for the assessment of process performance, control system evaluation and similar purposes is well established and the repeated reference to the first COST/IWA benchmark simulation model (BSM1) has clearly indicated the usefulness of such a tool for this purpose. However, the benchmark effort is not without its critics. The unit process sizes and model choices, influent characterisations, model transformations and evaluation criteria have all been criticised. Some of the criticism is justified as the defined BSMs have not always taken into account the most recent advancements or accepted theory, but the critics often fail to fully appreciate the benefits generated by the effort, the debates, the compromises and the solutions that have gotten the BSMs to this point. The BSM development group has ebbed and flowed over the years, but the group was typically made up of 20 or more modelling experts from around the world discussing the BSM development at regular meetings. What is further impressive is that this development has been carried out on a voluntary basis. The group, however, does recognise that the BSMs have limitations and even though this is consistent with any other mathematical models, it is important that benchmark users understand those limitations in order to assess the validity of their simulation studies. This chapter addresses those limitations.

9.1 BSM AS A TOOLBOX

About two decades ago, the concept of a tool that could be used to objectively evaluate the performance of control strategies through simulation using a standard model implementation was introduced for activated sludge wastewater treatment plants. That concept resulted in the development of BSM1, the subsequent BSM1_LT and most recently BSM2. Practitioners have suggested that these models are only academically applicable, have been conceived of for publication generation purposes and provide limited benefit to the applied modelling community. The authors of this STR beg to differ with those practitioners. It has further been suggested and it is recognised that control strategies that work well in a BSM may not work as well in practice due to the many factors affecting a full-scale plant. As authors of this STR, we are fully aware of this. Here, again it is important to keep in mind that the main purpose of the BSM development has been to create a tool for objective evaluation of control strategies through simulation. It has never been the primary plan to solve problems on a real plant with the BSMs. In addition to the successful use of BSMs for objective control strategy evaluation, the development of the BSM systems has provided a number of spin-off benefits and has resulted in the development of a number of universally applicable sub-models. The BSMs can be used in their entirety, but to limit their application this way would hide the value of the pieces that make up these models. Rather, the BSM systems also should be considered a modelling toolbox, and a platform on which modelling issues have been debated, experimented upon and tested.

The BSM sub-models should be considered a collection of modelling tools that address various aspects of whole-plant wastewater treatment modelling. The BSMs contain a whole series of these modelling tools including but not limited to: an influent wastewater generating model, a temperature model, empirical solids/liquid separation models, standard and verified implementations of ASM1, ADM1, the Takács double exponential settling model and the Otterpohl primary clarification model, anaerobic digestion/activated sludge model interfaces, a risk assessment model to infer solids separation problems of microbiological origin, performance indices, operational cost indices as well as models for energy consumption by aeration and pumping equipment and generic models for sensors and actuators. The BSM development value lies in these individual modelling tools, and the modular nature of those tools means that they are portable and can be used in isolation if the need arises. These tools have all appeared as a direct result of the BSM development, and without that effort, may not have appeared at all. Not only that, these tools have formed the basis for the development of standardised approaches to modelling issues like model interfaces and simulation speed optimisation as published beyond the BSM context (Nopens *et al.* 2009; Rosen *et al.* 2006; Volcke *et al.* 2006b). Some of the tools, (i.e., the ADM1 implementation (Technical Report No. 5) and the influent wastewater generation model (Gernaey *et al.* 2011)) are available as separate tools upon request, but all the tools resulting from the development are distributed for free and many come as standard components of the commercial WWTP simulators.

Model verification through ring-testing identified a number of implementation issues, coding errors and simulator specific insights, and it is quite likely that many of the identified issues would have remained unresolved in the commercial simulators had this effort not taken place. The ring-testing was instrumental in identifying minor problems under normal plant operating conditions. Extreme conditions were not tested, so it is unclear if other numerical issues still exist. However, one thing that was clearly illustrated through this process was the need to ring-test all new implementations. The Task Group was able to test the predominant software packages currently on the market, but undoubtedly new software packages will come to market in the future and users should ensure the quality of those software packages by testing them with the BSMs. Although it is comforting to assume that the software developers have implemented error-free code, the reality is that error-free code is rare, so caution is appropriate and the verification approach developed and applied in the BSM-work can help in minimising such errors in the future.

Users should also be cautioned that certain model implementations were heavily scrutinised and others were not tested as rigorously. The ring-testing effort was initiated to test the biokinetic model implementations and was later expanded to include the other unit process models in each software platform. The biokinetic models were extensively tested, but other models like the influent generator, or the calculation of the evaluation criteria were not evaluated to the same extent. As a result, users are cautioned to check the implementation of all sub-models in their platform of choice to ensure that simulator-specific issues will not affect the calculated results.

9.2 MODEL STRUCTURES

9.2.1 Biokinetic models

The BSMs are not all-encompassing tools nor are they 'best-practice' tools to be interpreted as showcasing the best models for specific unit processes. Many of the published models used in the BSMs were chosen based on compatibility and international acceptability at the time. The Task Group is not advocating the use of these models for purposes unrelated to benchmark studies as it is fully recognised that some of the benchmark models have specific short-comings.

For example, since its first introduction, ASM1 has been criticised because the published model does not contain ammonia switches in biomass growth rate expressions. Equally unfortunate is the omission of inorganic solids in the original description. Similarly, there is evidence suggesting that pH can affect the activated sludge activity, but pH is not considered in the published ASM1. The same holds true for processes such as enhanced biological phosphorus removal (EBPR), 2-step nitrification and denitrification including N_2O production and Anammox extensions to ASM1. Although these extensions and oversights have been included in many commercial simulator implementations, the benchmark description of ASM1 is the original published version.

The anaerobic digester in BSM2 is modelled using ADM1 (Batstone *et al.* 2002). However, it was necessary to implement a modified version of the published ADM1 to optimise the simulation performance and correct some mass balance issues (Rosen *et al.* 2006). Several changes were implemented including continuous inhibition functions for pH and an algebraic solution for the hydrogen state variable. The changes were the direct result of the BSM development and reflect the extensive investigation of ADM1 as it related to the simulation of the benchmark model. These changes represent significant improvements to the published model for the BSM2 implementation specifically and for practical simulation applications in general. Readers may want to adopt these changes for other non-BSM studies, but should do so while recognising that the changes were developed for BSM optimisation and may or may not be appropriate for every application of this model.

The inclusion (compared to BSM1) of primary treatment as well as sludge treatment in BSM2 increases the complexity of the system but more importantly allows for the study of unit process interactions. These inclusions have forced the benchmark team to develop and consider a new set of modelling tools that have far reaching implications and uses. Because the activated sludge and anaerobic digestion models each use different state variable sets, AS/AD interfaces were developed to account for a number of aspects including COD and nitrogen mass balances, charge continuity and differences in digester performance depending on the primary and secondary sludge composition of the digester influent. Here again, the interfaces were developed based on *a posteriori* knowledge of the system, the digester influent and the performance of the various unit processes. That is, a great deal of debate surrounding these interfaces was directed at the partitioning of the COD into appropriate state variables. A generalist approach was initially adopted, but this approach quickly identified performance issues that were deemed unrealistic in the context of BSM2. Based on this, the approach was modified with the goal of obtaining a particular result and the adopted model achieved that goal. As a result, the interface models may not be ideal for other applications. As with the other models, the Task Group believes that the adopted approach is applicable and that the model parameters are reasonable, but model users should realise that these parameters may not be applicable for all situations.

9.2.2 Aeration

Since first implemented into software tools, aeration models have been criticised for the simplifications that have been adopted. Many of the commercial simulators have updated their implementations and now include efficiency calculations in their transfer rates. That is, some of the more complicated models now attempt to predict variable transfer efficiencies based on the airflow and the airflow/diffuser. This is not the case for the BSM models. In the BSMs, aeration is modelled by directly varying the K_La. That is, if a DO controller calls for more oxygen, the model responds by increasing the K_La. The K_La is then translated into an estimated aeration energy quantity which is the output from the model. In reality the relationship between K_La, airflow and transfer efficiency is not that simple and experience tells us that blowers operating in an inefficient range can add significant costs to plant operations. So, users may need to consider this limitation if aeration control is a significant result from the control strategy that is being evaluated.

9.2.3 Solid/Liquid separation models

Several limitations related to the solid/liquid separation models are acknowledged. The thickening and dewatering units in BSM2 are assumed to be ideal. This is clearly not the case in reality but this assumption was deemed sufficiently accurate for BSM2 purposes. The Takács double exponential settling velocity model in a layer model is recognised as a reasonable representation of the settling process, but its inability to predict the impact of short term settling issues and compression makes this model unsuitable for judging absolute settling behaviour. That is, this model is ideally suited for examining relative changes in the predicted settling and sludge blanket behaviour and can be used to judge the impact that might result at a real facility, but the absolute changes in effluent solids predicted by the model may not be accurate. Furthermore, the adopted settling models do not include any biological conversions which may or may not be significant. The exclusion of biological activity in the clarifiers is a simplification that was deemed reasonable given the low sludge blankets in the BSMs, but different operating conditions could result in different settling conditions and it is widely recognised that denitrification in clarifiers can be significant, especially if a large blanket exists (e.g., Siegrist *et al.* 1995). Users should consider these limitations when performing benchmark simulations.

Settling problems can also be caused by a number of different things including hydraulics and microbial population shifts. These issues are not typically modelled and they are not included in the Takács model. However, it is widely recognised that bulking, foaming and rising sludge can be significant problems in reality. To address this issue, the BSMs have included a new risk analysis formulation that attempts to mathematically predict the propensity of a given treatment scheme to have settling problems (see section 4.5). The development of this approach is based on both the available literature and empirical knowledge used in practice, but the validity of the approach still needs to be evaluated with full-scale data. That is, issues like weir design that contribute to filamentous proliferation are not included in the model so it is quite possible that each model application will be unique.

9.2.4 Other models

Limitations exist in the other models as well. Pumps, for instance, are assumed to be ideal in the basic BSM implementations. That is, if a pump set point is changed, the pumped flow is assumed to change instantaneously without error to the new rate unless an actuator model is used to control the change. This simplification has the potential to generate unrealistic output as real pumps do not operate in this way, nor do they pump precisely constant flows over extended periods of time. Care should be taken by the BSM user if the control strategy under study relies heavily on a pumped flow control scenario. The sensor and actuator models that were developed as part of the BSM1_LT work allow including this more realistic behaviour, dynamic response to changes, delays and drift so as to mimic reality better. They come at a cost, however, as more implementation work is required and they typically result in slower computations.

The modelling of sensors is slightly more detailed as signal response times are modelled. It was reasoned that if a control strategy required a sensor and the sensor signal delay was significant, then the evaluation should take this into account. For example, the evaluation of a respirometry-based control strategy would certainly be unrealistic if the OUR measurement on which it was based was assumed to be instantaneous when in reality the actual measurement might take 10, 20, 30 minutes or longer to acquire. However, as the BSMs were developed so that control strategies could be compared, it was necessary to limit the permutations that a user could choose when modelling a sensor. Much like the reasoning behind choosing the biological parameters, it was not appropriate (as the results would have been too hard to compare) to allow users the option to choose all of the parameters for a particular sensor without limiting those choices in some way. In this case, sensor classes with specific features were defined to limit those choices (Rieger *et al.* 2003). This is, therefore, by definition a limitation and the user should recognise that it is more than possible that these classes will not be explicitly consistent with a particular sensor. As a result, users will be forced to choose a sensor class that has characteristics that are the closest match to their sensor. This limitation may skew the results slightly, but by limiting the sensor definitions, the results of any particular BSM study will be more readily comparable to other studies.

Finally, the way the temperature dynamics in the bioreactors is modelled in BSM1_LT and BSM2 is kept very simple. The main temperature dynamics in the influent follow a sinusoidal variation over the year, which may or may not be accurate and the temperature in combined streams is based on a flow-weighting equation. No heat generation or losses are accounted for. Furthermore, BSM2 assumes a constant temperature in the anaerobic digester which may or may not be significant for some strategies.

9.3 MODEL PARAMETERS

The model parameters defined in the BSMs are for control evaluation purposes only. The chosen parameters are believed to be reasonable for all of the unit processes, but the Task Group is not suggesting that these parameters be used for any other modelling purpose. That is, each of the sub-models has a number of defined parameters. These parameters have been chosen for the BSMs to achieve certain goals like removal efficiencies or are based on certain assumptions like temperature (e.g., 15°C for BSM1). The Task Group makes no claim as to the validity of those parameter value choices, nor does the Task Group purport that the assumptions on which those parameters are based are correct. The Task Group believes those assumptions to be reasonable, but the choice of model parameters outside the benchmark environment should be based on sound, local knowledge related to the modelling task at hand.

For example, BSM1 assumes a temperature of 15°C and the defined parameter values in that system reflect parameter estimates at that temperature. BSM1_LT and BSM2 have a variable temperature so temperature corrections based on a temperature correction equation have been adopted for the kinetic parameters. These corrections have been assumed on the basis of literature information so they may not be the best estimates for those parameters at any given temperature. The digester is assumed to have a constant temperature, but the activated sludge portion of the model has a variable temperature, mainly driven by the influent temperature. The Task Group believes that these are reasonable assumptions and simplifications, but the reader should not deem these as the best or only approach for other applications. Any temperature modelling done outside of the BSMs should consider local and appropriate assumptions where necessary, but the approach adopted in the BSM development may provide a reasonable place to start in the absence of more detailed local information.

9.4 EVALUATION CRITERIA

The development of the evaluation criteria required a significant amount of discussion, debate and compromise. Through this process, the Task Group and various contributing research groups were enlightened as to the many different effluent regulations that exist across the globe. Some groups advocated for mass load limits, while others wanted every gram of effluent pollution counted. Some groups wanted 'soft' limits like 95 percentiles and others were content with firm concentration limits. As the goal was to develop criteria that were location independent, it was agreed that the use of concentration limits was the best approach. However, here again, it was not straightforward as to what those concentration limits should be. In the end, the limits were arrived at through a debate and compromise and are not meant to reflect limits in a particular location.

Similar discussions surrounded the other criteria as well. The EQI factors are loosely based on Flemish legislation (Vanrolleghem *et al.* 1996) and the energy calculations are the result of various proposals. For example, because the options for real pumping equipment are considerable and the characteristics of that equipment vary, it was necessary to make a number of assumptions to achieve a pumping energy number for the OCI calculation. This process was repeated for the aeration, mixing and heating energy calculations. As a result, it is important to note that the calculated quantities may not be completely accurate, but instead should be considered to be relative terms. That is, if the OCI increases from one scenario to another, it is presumed that the operating costs will rise if those changes were made at full scale. The absolute increase in operating costs may not be known, but the relative comparison of the costs suggests that an increase is likely. A similar conclusion can be reached about changes in the calculated EQI. An increase in the EQI suggests that, relatively speaking, the effluent quality will deteriorate.

A significant amount of debate during development surrounded capital cost evaluations. The implementation of control is expected to require a capital cost allocation. Some plants will already have the infrastructure in place, but most plants will not. As capital costs have not been included, it is possible that a control strategy with an unrealistic capital expenditure could be developed. This strategy could result in an evaluation criteria output that is exceptional without being penalised for the capital costs. This is a limitation that users should be aware of. In reality, the gain in OCI of the control strategy might be compared to the capital expenditures, for instance in terms of a payback period or an internal rate of return, but this would depend on the local situation. Indeed, the major problem with estimating capital costs is that they are often location specific and a conscious effort has been made to avoid location specific criteria. This results in a conflict that has been addressed qualitatively rather than quantitatively. When evaluating a control strategy, it is recommended that users include a list of new equipment (sensors, actuators, pumps, tanks, pipes,...) that is required for the strategy. This list may not be comparable to the other evaluation criteria directly, but it will make the comparison of a strategy that reports a modest operational cost index gain with an extensive equipment requirement list comparable to another strategy with limited infrastructure requirements. It is important always to keep in mind that the evaluation criteria represent relative changes and are simply indicative of potential differences between one strategy and another and this list achieves that goal.

9.5 MODEL SIMULATION

The evaluation criteria in all of the BSMs are based on an arbitrary data resolution of 15 minutes. This frequency provides sufficient resolution to ensure that the simulation dynamics will not be missed. It was reasoned that if the output was recorded less frequently, underlying plant dynamics could occur in between sampling times in which case the simulation output could miss an event completely. The opposite is also true as less frequent output could result in peak measurements only and miss the valleys. More frequent output (say at 1 minute intervals) was considered as well, but ultimately this was rejected because the added data was thought to provide little if any benefit to the dynamic evaluation. Using a 15 minutes data resolution was deemed to be a reasonable compromise between the amount of data output, simulation speed and event resolution.

BSM1 was developed using a 7-day evaluation period. From a control perspective this may well be sufficient if the control loop is fast and the impact of the control action is equally rapid. That is, aeration and recycle pumping control loops have sufficiently short time constants to be analysed over a 7-day period. When a dissolved oxygen control loop calls for more air, the blowers ramp up and more air is supplied within minutes. If a recycle pump is turned up, the impact on the process can be seen almost immediately in the simulation output, even though the full impact on the sludge concentrations follows the hydraulic retention time. However, three hydraulic retention times fall still within the 7-day period. So, for fast control loops like these, 7 days may well be sufficient.

However, a limitation with this time period presents itself when control loops with longer time constants are evaluated. For example, control strategies that aim to control the solids retention time will need far more time than 7 days of evaluation because the impact of changing the waste sludge flow rate may take more than 10 days to materialise in the simulation output. Seven days is equally problematic for evaluating stochastic events like sensor or actuator faults. Different types of faults will have different and time varying impacts and 7 days may not be sufficient to observe those effects. Modelling temperature impacts cannot be done in a 7 day period either. The liquid temperature in a treatment plant varies gradually over time and may be at its highest or lowest for only a short period of time over the course of the year. So, it is unrealistic to model larger temperature changes over that short a period and it may be equally unrealistic to model a low temperature steady state solution as a real plant never actually experiences that temperature long enough to reach steady state. To address this issue, BSM1_LT and BSM2 have adopted a one year (52 weeks, 364 days) evaluation period. Using a period of this length provides the BSM user with the option to evaluate more control strategy options that could not be effectively evaluated using 7 days and observe the effect of seasonality in temperature and precipitation. Regardless of the evaluation time being used, the BSM user should be aware of the limitations introduced by the length of these evaluation periods.

9.6 APPLICATION EXTENSION

The use of the modelling toolbox provides an excellent starting point for modelling and evaluating many systems. Examples of such applications have been presented by Abusam (2001), Volcke *et al.* (2006a) and Benedetti *et al.* (2006). The ability to 'extend' the BSMs is an unquantifiable benefit of the work and shows that the BSM influence should not be restricted to control strategy evaluation alone. Rather, this work has illustrated its importance and associated benefits to modelling in general.

The toolbox has been freely distributed to research groups on all continents to provide a structured, documented and validated starting point for their future work. The development of the BSM systems has also provided a number of spin-off benefits beyond the models themselves. This development effort has been criticised in some circles as having limited value due to the theoretical approach and the fact that the model is not based on any particular plant. However, these models were developed as a means to develop and test control strategies in an unbiased fashion. That is, although there are numerous literature studies on the effectiveness of various control strategies, it is typically not possible to compare these studies because the conditions used to test the strategies are different. To address this issue, the descriptions of the BSM simulation procedures are rigid to ensure the consistent application of the benchmark and to ensure that similar analyses are done on the output data for each simulated control strategy. Only with a completely defined simulation procedure can any conclusions be reached when one strategy is compared to another. So, although some of the models already existed, the protocol for the standardisation of the output evaluation is an interesting development. The evaluation criteria are not exhaustive, incorporate a number of limitations and have many inherent assumptions, but they do represent a standardised approach.

The BSM definitions discussed in this STR are rigidly based on a pre-denitrifying activated sludge system in a Modified Ludzack-Ettinger (MLE) configuration (this is the default set-up of the BSMs although available control handles allow for post-denitrification systems, various step-feed systems, anox-aerobic-anox-aerobic systems and many others). Because of this rigidity in the definition, the application of the BSMs, that is, the evaluation of control strategies, is somewhat limited to plants with this configuration, which is only a subset of the plants in operation. During the initial stages of the BSM development, Copp *et al.* (2002) expanded the COST benchmark work (Copp, 2002) to include nitrifying and carbon removal only configurations. Although the models developed in that work are not entirely consistent with the BSM described here, it does show that portions of the BSMs can be modified for specific applications if the need arises. Extensions have also been made within the BSM framework to evaluate oxidation ditches (Abusam, 2001; Abusam *et al.* 2002), Sharon (Volcke, 2006), Anammox (Volcke *et al.* 2006a) and MBR systems (Maere *et al.* 2011). These BSMs are fully defined, but as the application of the framework expands, it is fully expected that new extensions and configurations will be developed for other processes such as EBPR (Gernaey & Jørgensen, 2004), simultaneous nitrification/denitrification and greenhouse gas emissions (Flores-Alsina *et al.* 2011; Guo & Vanrolleghem, 2014) and the fate of micro-pollutants (Snip *et al.* 2014). Although, defining the configuration is a limitation, the modular nature of the framework allows for future development and should enable continued use of the modelling toolbox.

9.7 CONCLUSION

The BSM development has been an international collaborative effort that has resulted in a number of new model initiatives. This effort has provided significant insights into the modelling of wastewater treatment from a mathematical and numerical point of view. It has set the scene for proper implementation procedures and ring-testing schemes. It has produced a number of new models (for influent, sensors, error, interfacing and evaluation) or important modifications (ADM1) and has provided standardised simulation protocols. It is critically important to realise that none of these models have been calibrated to real data from a full-scale facility (within the framework of the BSM development) and that the processes being modelled are not based on a real plant. However, it is equally important to point out that the BSMs represent years of discussions, debates and compromises by a comprehensive, global expert team and although not perfect they have identified and focused attention on numerous modelling issues that did not have standardised solutions prior to this work.

Chapter 10

Conclusions and perspectives

K. V. Gernaey, J. B. Copp, U. Jeppsson and P. A. Vanrolleghem

This Scientific and Technical Report (STR) summarises the result of approximately 20 years of development work on the different benchmark platforms. Apart from the benchmark platforms themselves, this Task Group has been able to produce a set of verified and platform independent unit process models that are essential for WWTP control strategy evaluation simulation studies. Furthermore, the Task Group has spearheaded the development of a set of generic simulation tools that can be used to support the modelling of activated sludge WWTPs in general. In this chapter, conclusions are drawn on the basis of our experience with the development and the usage of the benchmark platforms and an outlook towards future developments is provided.

10.1 LESSONS LEARNED: DEVELOPMENT OF THE BENCHMARK PLATFORMS

As discussed throughout this STR, the benchmark platforms – BSM1, BSM1_LT and BSM2 – are comprehensive tools and freely available simulation protocols which allow for the objective comparison of WWTP control and/or monitoring strategies.

It is important to emphasise that the development of the benchmark platforms was a voluntary effort for the most part. Considering the large number of scientific papers that have appeared on the development and the use of the benchmark platforms, one important lesson learned is that voluntary work still has a major role to play in the scientific community. Of course, there has been some financial support for meetings and the Task Group is grateful for the support from IWA and the COST actions 682 and 624. Clearly, the development of the benchmark platforms has demonstrated that such support (e.g., COST actions) is important for promoting frequent contact between researchers working in similar areas. This development would not have been possible without that support. Regular face-to-face discussions are indeed essential when embarking on this type of journey and these meetings also facilitated the interaction of MSc and PhD students whose projects were modulated to fit into the benchmark development project.

Another lesson learned from this development is that collaboration between industry and academia is essential for a successful outcome. Indeed, the fact that the development has been supported by a broad and diverse group of people is key to its success in the wastewater modelling and simulation community, and this has been assisted by the fact that the development, from the outset, has been simulation platform independent.

The benchmark development also has resulted in many spin-off benefits. The first indication of these benefits was the ASM1 ring-testing – the confirmation of results using different software platforms and different developers – that was carried out very early on in the BSM1 development. Obscure errors were discovered and simulator-specific issues were identified. Without the rigorous testing that BSM1 required, it is not unreasonable to conclude that those errors might still exist in the commercially available software packages. Ring-testing of all the unit process models has, in general, been a very time-consuming exercise, and has been one of the most challenging tasks, but it has provided some of the most important spin-off benefits. Further benefits became apparent during the BSM2 development because a significant amount of time was spent on model development issues such as the ADM1 implementation, an influent wastewater generator model, ASM1 → ADM1 and ADM1 → ASM1 model interfaces, in addition to several others. The effort put into these issues is important because they are generic tools that might be (and already are) used outside of the benchmark environment.

The expansion of BSM1 to include primary clarification and sludge treatment with anaerobic digestion for BSM2 resulted in some interesting developments from a control point of view. An initial draft of BSM2 with the activated sludge tank volumes identical to the BSM1 volumes was presented at WaterMatex2007 (Jeppsson *et al.* 2007). At that time, it was shown using 15 simple control strategies that control of that plant was limited and control authority was insufficient to significantly improve the performance of the highly overloaded plant layout. Interestingly, the high nitrogen load that was causing some of the issues was associated with the reject water, which was not present or accounted for in the BSM1 case. The activated sludge tanks in BSM2 were thus redesigned and the current layout was presented several years later by Nopens *et al.* (2010). This layout (compared to the earlier versions) included: (1) a reduced influent nitrogen load to compensate for the reject water contribution; (2) increased activated sludge tank volumes, compared to BSM1, in order to obtain a WWTP that can benefit from process control; and (3) modifications to the evaluation criteria to distinguish between nitrate and ammonia nitrogen in the effluent quality index, to reduce the aeration impact on the calculated cost index and to differentiate the energy consumption for different pumped flows.

10.2 LESSONS LEARNED: USE OF THE BENCHMARK PLATFORMS, VERIFIED PROCESS MODELS AND GENERIC TOOLS

The benchmark platforms have been well received by the research community. BSM1 and BSM2 are used frequently for their original purpose (e.g., Vrecko *et al.* 2002; Stare *et al.* 2007). They are relatively easy to use because the different parts (unit process models, control strategies, evaluation criteria,…) are well documented and tend to be immediately recognisable by most users. The concept of benchmarking monitoring strategies, as introduced in BSM1_LT, has been used less frequently thus far, but this benchmark provides a necessary platform nevertheless in a niche area of research. The Task Group hopes that BSM1_LT will receive more interest and be a source of inspiration in the future, as monitoring strategies and methods become increasingly important for WWTPs.

10.2.1 Portability

The modular construction of the BSM platform is one of the main reasons for its success. Because the benchmark can be viewed as being the combination of several generic tools, it is relatively easy for a user to pick one or more of these tools and use them in isolation or as part of another study. Furthermore, this structure gives users the ability to add on one or more additional features to the BSM platform without the need to implement a significant amount of computer code. This is illustrated with some examples below.

Quite recently, Sin *et al.* (2009, 2011) used the BSM1 plant layout to demonstrate the application of sensitivity and uncertainty analysis methods. This application was not envisioned when the benchmark development was initially discussed, but is an interesting illustration of how the BSM platform has been used as a convenient vehicle for the demonstration of new ideas or tools.

In other cases, benchmark users have not used the BSM layouts, but have instead used one of the benchmark tools (e.g., the evaluation criteria) to compare control strategy performance on models of other plants. Abusam *et al.* (2002) was one of the first to apply the BSM1 evaluation criteria to another plant model. Many similar applications of the evaluation criteria can be found in the literature. Other authors have used the BSM1 influent data files (dry, storm and rain weather data) or transformations thereof in order to simulate a reasonable dynamic influent.

The Task Group is pleased that users take parts of the BSM platform, and use these in their research projects. However, it is critical that users clearly document what parts they have used and what transformations they have implemented. Unfortunately this is not always the case and the documentation in some studies is insufficient. It is important that the users specify precisely which parts of the BSM platform have been used and also provide a detailed description of any additions. This is essential to ensure that other users understand what has been done and can duplicate the work if necessary.

10.2.2 Extensions

With the extension of BSM1 to the plant-wide BSM2, the number of potential extensions has increased substantially. Volcke *et al.* (2006a) were the first to modify the BSM2 layout with reject water treatment. More recently, the Task Group has observed quite a number of publications where the BSM2 plant layout was extended with extra functionality. For example, one known limitation of the BSM biological model is that only one-step nitrification and denitrification are used, but there has been a growing interest in predicting greenhouse gas formation from the treatment process, and N_2O – an important nitrification and denitrification intermediate with a high greenhouse gas (GHG) potential – is one of the compounds that is being studied extensively. Multi-step nitrification/denitrification models including N_2O production have already been implemented in the BSM2 benchmark framework (Flores-Alsina *et al.* 2011; Corominas *et al.* 2012; Guo *et al.* 2012; Guo & Vanrolleghem, 2014). Several other potential extensions have been described by Jeppsson *et al.* (2013). In many cases these add-ons have been defined, implemented and remain the property of individual research groups which contrasts the Task Group's philosophy that verified implementations of the benchmark plants are to be freely distributed. Clearly, the Task Group supports such extensions. It is certainly hoped that these extensions will be made available in the future, and that the documentation of such extensions will be extensive so that the entire research community can benefit. The Task Group invites all interested parties to develop additional tools, but emphasises that model extensions should be ring-tested (e.g., by comparison of two or three independent implementations by different users) before being distributed. Ring-testing these extensions is the best way to ensure the quality of the documentation and the computer code. Although not currently ready, the Task Group is discussing the definition of a suitable protocol to ensure the quality of any additions.

10.3 LOOKING AHEAD: FUTURE EXTENSIONS OF THE BSM PLATFORMS

One major conclusion is that the benchmarking work is far from finished. A wish list for future extensions to the BSM platforms was presented in June 2011 at the Watermatex conference in San Sebastian (Spain) by the Task Group chairman, and this work was later published in a journal paper (Jeppsson *et al.* 2013). Potential extensions can be subdivided into several categories including: (1) Temporal extensions; (2) Spatial extensions; (3) Process extensions within the WWTP; (4) Realism of the models used in the BSM; (5) Control strategy extensions; and (6) Extended evaluation tools. The most important trends are briefly highlighted here.

10.3.1 Temporal extensions

Dynamic simulations with BSM1 were based on 14-day influent files depicting dry, rain and storm conditions, but this approach was shown to be insufficient for long-term control applications such as sludge age effects and equipment failure scenarios. To address this shortcoming, a phenomenological influent wastewater generator model was developed to cope with the challenge of providing realistic long-term (1–2 years) influent data to the BSMs. Even longer influent data files might be required for future scenario analyses. For example, the simulation of changes in urban catchments and the impact on treatment plant efficiency might require evaluation periods in excess of ten years. The ever-increasing computational power would allow an extended evaluation period; however, the practical relevance of results based on such extended simulations would likely be highly uncertain. Nevertheless, such an extension may be well-received in the future.

10.3.2 Spatial extensions

The benchmark platforms described in this STR are defined as 'within-the-fence' systems, that is, the model descriptions and simulations do not extend outside the borders of the WWTP. The importance of the sewer system and processes in the receiving waters were recognised by the Task Group but including these complicating factors in the original BSMs was deemed to be beyond the Task Group's scope. From a control and monitoring perspective the inclusion of the sewer network into the benchmark system would open up a range of new possibilities for interactions and manipulation of the combined sewer/WWTP system (e.g., back-flow effects, storm tanks and pumping stations, combined sewer overflows, pollution contributions from run-off). Examples of integrated sewer and WWTP modelling can be found in Sharma *et al.* (2012) and an initial combined benchmarking study connecting the sewer model with BSM2 was performed by Guo *et al.* (2012). As for receiving waters, existing models such as the River Water Quality Model No. 1 (Reichert *et al.* 2001) could be added or linked to the BSMs without too much difficulty given that the model interfaces have already been developed (Benedetti *et al.* 2007). This kind of approach would be particularly beneficial for a more detailed evaluation of the environmental impact of wastewater pollutants. Moreover, such a combination would promote the use of the benchmark system as a decision support tool in agreement with current river basin management approaches, as pursued for example by the EU Water Framework Directive.

Integrated evaluation experience can be found in Benedetti *et al.* (2010) and Brehmer *et al.* (2009). As with the benchmarking tools developed so far, consensus will have to be reached on objective evaluation criteria that assess the urban water quality impacts in receiving waters, but ideas for this are being developed (e.g., Bauwens *et al.* 1996; Vanrolleghem *et al.* 2005a; Benedetti *et al.* 2010; Clouzot *et al.* 2013).

10.3.3 Process extensions

Process extensions within the WWTP are related to the appearance of models for new unit processes, such as the SHARON and ANaerobic AMMonium OXidation (Anammox) processes (Volcke *et al.* 2006a). The Task Group expects that this evolution will continue in the future. The Task Group has recognised an immediate need for process extensions related to integrated fixed-film processes (Vanhooren *et al.* 2002) and SBR configurations. Note that these process extensions will have to be coupled to suitable model interfaces if the state variables in the new model differ from the state variables in the original benchmark models. As discussed previously in this STR, these interfaces ensure that material mass balances and continuity principles are met when mapping the output variables of one model to the most appropriate input variables of another model (Alex *et al.* 2005; Vanrolleghem *et al.* 2005b; Volcke *et al.* 2006b; Nopens *et al.* 2009) and these issues are critical for acceptance within the modelling community.

10.3.4 Realism of models used in BSM

The mathematical models used in the benchmark platforms were chosen because they are internationally accepted and well established. The Task Group is aware of the limitations in the chosen models; but, is also keenly aware of the fact that there is an almost limitless possibility to extend and upgrade the models within the existing BSM plant configurations, including the models describing sensors and actuators (e.g., more detailed and dynamic models of blowers and pumps). It is unequivocally hoped that the aim of any changes would be to enhance realism of the systems rather than to simply increase the level of detail and complexity. In some cases, improved models have become available and although the Task Group decided not to change models during the development phase, most of these updated models are well described, and could be interchanged within the current BSM framework. For example, future extensions may include phosphorus removal, multi-step nitrification/denitrification and N_2O production, inclusion of sulphur reducing/oxidising reactions, improved thickener and dewatering models, improved settler models, inclusion of soluble microbial products formation and extensions for physico-chemical processes. Furthermore, the BSM structure provides a framework to test models under development (e.g., Guo & Vanrolleghem, 2014).

10.3.5 Control strategy extensions

Advances in instrumentation and automation allow us to have access to highly accurate information regarding the urban wastewater system in real-time. Not only can this information be acquired from the WWTP, but on-line sensors can be installed in the sewer systems, and monitoring stations are being developed for monitoring river water quality (Copp *et al.* 2010). These developments are producing large quantities of data that can be used for fault-tolerant, uncertainty-aware and system-wide control design (Olsson *et al.* 2005; Olsson, 2012). Certainly, the benchmark platforms will be used in the future to test such new control strategies and the Task Group hopes that these new trends in process control, applied to the benchmark platforms, might contribute to our understanding on how such new control strategies can contribute to improving the operation of full-scale WWTPs.

10.3.6 Extended evaluation tools

The performance evaluation in the current benchmark platforms is based on three main types of evaluation criteria (effluent quality, operational cost issues and process risks). Effluent quality is considered through an effluent quality index (EQI), costs for energy, sludge and chemicals are considered through an operating cost index (OCI) and process risk is considered through a fuzzy logic calculation of microbiology-related operational problems to create a risk index. The wish list for future expansions of the evaluation criteria contains energy consumption models, validation and extension of the risk calculation concerning microbiology-related TSS separation problems, inclusion of capital and maintenance costs, robust uncertainty based evaluation of control strategies (e.g., the robustness index proposed by Vanrolleghem &

Gillot, 2002), as well as a set of criteria to capture greenhouse gas emissions and the carbon footprint of the WWTP (e.g., the proposal discussed in Corominas *et al.* 2012).

10.4 THE 'BENCHMARKING SPIRIT'

Two central elements in the benchmarking community philosophy have been the development of validated models and the free distribution of the verified models. All of this has been possible because of a tremendous amount of voluntary work. With over 400 scientific papers published over the years, the results of that voluntary effort are very impressive.

It is the Task Group's sincere hope that this report will inspire research groups to carry on the development effort allowing it to flourish and continue to be a state-of-the-art tool for research, model development and practical application of control within the field of wastewater treatment. This STR is the culmination of this stage in the BSM development, but with so many possibilities for expansion and improvement, hopefully a new wave of motivated scientists will foster new like-minded life-long friendships and continue the development well into the future.

References

Abusam A. (2001). Development of a Benchmarking Methodology for Evaluating Oxidation Ditch Control Strategies. PhD thesis, Wageningen University, The Netherlands, ISBN 90-5808-422-1.

Abusam A., Keesman K. J., Spanjers H., van Straten G. and Meinema K. (2002). Evaluation of control strategies using an oxidation ditch benchmark. *Water Science and Technology*, **45**(4–5), 151–158.

Alex J., Beteau J. F., Copp J. B., Hellinga C., Jeppsson U., Marsili-Libelli S., Pons M. N., Spanjers H. and Vanhooren H. (1999). Benchmark for evaluating control strategies in wastewater treatment plants. European Control Conference (ECC'99), Karlsruhe, Germany, August 31–September 3, 1999.

Alex J., Ogurek M. and Jumar U. (2005). Formalised model presentation for wastewater systems. 16th IFAC World Congress, Prague, Czech Republic, July 3–8, 2005.

ASCE (1993). A Standard for the Measurement of Oxygen Transfer in Clean Water. ASCE Transfer Standard Committee, New York, NY, USA.

ATV A131 (2000). ATV-DVWK Arbeitsblatt A 131: Bemessungen von einstufigen Belebungsanlagen ab 5000 EW. ATV, Hennef, Germany (in German).

Batstone D. J., Keller J., Angelidaki I., Kalyuzhnyi S. V., Pavlostathis S. G., Rozzi A., Sanders W. T. M., Siegrist H. and Vavilin V. A. (2002). Anaerobic Digestion Model No. 1 (ADM1), IWA Scientific and Technical Report No. 13, IWA Publishing, London, UK.

Bauwens W., Vanrolleghem P. A. and Smeets M. (1996). An evaluation of the efficiency of the combined sewer – wastewater treatment system under transient conditions. *Water Science and Technology*, **33**(2), 199–208.

Benedetti L., Bixio D. and Vanrolleghem P. A. (2006). Benchmarking of WWTP design by assessing costs, effluent quality and process variability. *Water Science and Technology*, **54**(10), 95–102.

Benedetti L., Meirlaen J., Sforzi F., Facchi A., Gandolfi C. and Vanrolleghem P. A. (2007). Dynamic integrated water quality modelling: a case study of the Lambro river, northern Italy. *Water SA*, **33**(5), 627–632.

Benedetti L., De Keyser W., Nopens I. and Vanrolleghem P. A. (2010). Probabilistic modelling and evaluation of waste water treatment plant upgrades in a water quality based evaluation context. *Journal of Hydroinformatics*, **12**(4), 380–395.

Brehmer I., Reußner F., Schütze M., Muschalla D. and Ostrowski M. (2009). Weiterentwicklung des hessischen Leitfadens zum Erkennen ökologisch kritischer Gewässerbelastungen durch Abwassereinleitungen – Entwicklung einer simulationsgestützten Planungsmethodik. *Korrespondenz Abwasser*, **56**(4), 382–384 (in German).

Bürger R., Diehl S. and Nopens I. (2011). A consistent modelling methodology for secondary settling tanks in wastewater treatment. *Water Research*, **45**(6), 2247–2260.

Chen J. and Beck M. B. (1993). Modelling, control and on-line estimation of activated sludge bulking. *Water Science and Technology*, **28**(11–12), 249–256.

Clouzot L., Choubert J.-M., Cloutier F., Goel R., Love N. G., Melcer H., Ort C., Patureau D., Plósz B. G., Pomiès M. and Vanrolleghem P. A. (2013). Perspectives on modelling micropollutants in wastewater treatment plants. *Water Science and Technology*, **68**(2), 448–461.

Comas J., Rodríguez-Roda I., Sànchez-Marrè M., Cortés U., Freixó A., Arráez J. and Poch M. (2003). A knowledge-based approach to the deflocculation problem: Integrating on-line, off-line, and heuristic information. *Water Research*, **37**(10), 2377–2387.

Comas J., Rodríguez-Roda I., Gernaey K. V., Rosen C., Jeppsson U. and Poch M. (2008). Risk assessment modelling of microbiology-related solids separation problems in activated sludge systems. *Environmental Modelling and Software*, **23**(10–11), 1250–1261.

Copp J. B. (1999). Development of standardised influent files for the evaluation of activated sludge control strategies. In: Respirometry in Control of the Activated Sludge Process: Benchmarking Control Strategies, J. B. Copp, H. Spanjers and P. A. Vanrolleghem (eds) (2002), IWA Scientific and Technical Report No. 11, IWA Publishing, London UK. ISBN 19-002-2251-5 – internal Task Group report.

Copp J. B. (ed.) (2002). The COST Simulation Benchmark – Description and Simulator Manual. Office for Official Publications of the European Communities, Luxembourg, ISBN 92-894-1658-0.

Copp J. B., Spanjers H. and Vanrolleghem P. A. (eds) (2002). Respirometry in Control of the Activated Sludge Process: Benchmarking Control Strategies, IWA Scientific and Technical Report No. 11, IWA Publishing, London, UK.

Copp J. B., Jeppsson U. and Rosen C. (2003). Towards an ASM1 – ADM1 state variable interface for plant-wide wastewater treatment modeling. 76th Annual WEF Conference and Exposition (WEFTEC), Los Angeles, CA, USA, October 11–15, 2003.

Copp J. B., Belia E., Hübner C., Thron M., Vanrolleghem P. A. and Rieger L. (2010). Towards the automation of water quality monitoring networks. Automation Science and Engineering (CASE), Toronto, Ontario, Canada, August 21–24, 2010.

Corominas L., Villez K., Aguado D., Rieger L., Rosen C. and Vanrolleghem P. A. (2011). Performance evaluation of fault detection methods for wastewater treatment processes. *Biotechnology and Bioengineering*, **108**(2), 333–334.

Corominas L., Flores-Alsina X., Snip L. and Vanrolleghem P. A. (2012). Comparison of different modelling approaches to better understand and minimize greenhouse gas emissions from wastewater treatment plants. *Biotechnology and Bioengineering*, **109**(11), 2854–2863.

Downs J. J. and Vogel E. F. (1993). A plant-wide industrial process control problem. *Computational and Chemical Engineering*, **17**(3), 245–255.

Ekama G. A., Sotemann S. W. and Wentzel M. C. (2007). Biodegradability of activated sludge organics under anaerobic conditions. *Water Research*, **41**(1), 244–252.

Fiter M., Güell D., Comas J., Colprim J., Poch M. and Rodríguez-Roda I. (2005). Energy saving in a wastewater treatment process: an application of fuzzy logic control. *Environmental Technology*, **26**(11), 1263–1270.

Flores-Alsina X., Comas J., Rodriguez-Roda I., Gernaey K. V. and Rosen C. (2009). Including the effects of filamentous bulking sludge during the simulation of wastewater treatment plants using a risk assessment model. *Water Research*, **43**(18), 4527–4538.

Flores-Alsina X., Corominas Ll., Snip L. and Vanrolleghem P. A. (2011). Including greenhouse gas emissions during benchmarking of wastewater treatment plant control strategies. *Water Research*, **45**(16), 4700–4710.

Gernaey K. V. and Jørgensen S. B. (2004). Benchmarking combined biological phosphorus and nitrogen removal wastewater treatment processes. *Control Engineering Practice*, **12**(3), 357–373.

Gernaey K. V., Rosen C., Benedetti L. and Jeppsson U. (2005). Phenomenological modelling of wastewater treatment plant influent disturbance scenarios. 10th International Conference on Urban Drainage (10ICUD), Copenhagen, Denmark, August 21–26, 2005.

Gernaey K. V., Rosen C. and Jeppsson U. (2006). WWTP dynamic disturbance modelling – an essential module for long-term benchmarking development. *Water Science and Technology*, **53**(4–5), 225–234.

Gernaey K. V., Flores-Alsina X., Rosen C., Benedetti L. and Jeppsson U. (2011). Dynamic influent pollutant disturbance scenarios generation using a phenomenological modelling approach. *Environmental Modelling and Software*, **26**(11), 1255–1267.

Gossett J. M. and Belser R. L. (1982). Anaerobic digestion of waste activated sludge. *Journal of the Environmental Engineering Division ASCE*, **108**(EE6), 1101–1120.

Guo L., Porro J., Sharma K. R., Amerlinck Y., Benedetti L., Nopens I., Shaw A., Van Hulle S. W. H., Yuan Z. and Vanrolleghem P. A. (2012). Towards a benchmarking tool for minimizing wastewater utility greenhouse gas footprints. *Water Science and Technology*, **66**(11), 2483–2495.

Guo L. and Vanrolleghem P. A. (2014). Calibration and validation of an Activated Sludge Model for Greenhouse gases no. 1 (ASMG1): Prediction of temperature-dependent N_2O emission dynamics. *Bioprocess Biosystems Engineering*, **37**(2), 151–163.

Hauduc H., Rieger L., Takács I., Héduit A., Vanrolleghem P. A. and Gillot S. (2010). A systematic approach for model verification: Application on seven published activated sludge models. *Water Science and Technology*, **61**(4), 825–839.

Harris R. W., Cullinane M. J. and Sun P. (eds) (1982). Process Design & Cost Estimating Algorithms for the Computer Assisted Procedure for Design and Evaluation of Wastewater Treatment Systems (CAPDET). US EPA, Office of Water Program Operations, USA.

Henze M., Grady Jr. C. P. L., Gujer W., Marais G. v. R. and Matsuo T. (1987). Activated Sludge Model No. 1, IAWPRC Scientific and Technical Report No. 1, IWA Publishing, London, UK.

Huete E., de Gracia M., Ayesa E. and Garcia-Heras J. L. (2006). ADM1-based methodology for the characterisation of the influent sludge in anaerobic reactors. *Water Science and Technology*, **54**(4), 157–166.

Hug T., Gujer W. and Siegrist H. (2006). Modelling seasonal dynamics of Microthrix parvicella. *Water Science and Technology*, **54**(1), 189–198.

Jenkins D., Richard M. G. and Daigger G. T. (2003). Manual on the Causes and Control of Activated Sludge Bulking, Foaming & other Solids Separation Problems, 3rd edn., IWA Publishing, London, UK.

Jeppsson U., Rosen C., Alex J., Copp J., Gernaey K. V., Pons M.-N. and Vanrolleghem P. A. (2006). Towards a benchmark simulation model for plant-wide control strategy performance evaluation of WWTPs. *Water Science and Technology*, **53**(1), 287–295.

Jeppsson U., Pons M.-N., Nopens I., Alex J., Copp J., Gernaey K. V., Rosen C., Steyer J.-P. and Vanrolleghem P. A. (2007). Benchmark Simulation Model No 2 – General protocol and exploratory case studies. *Water Science and Technology*, **56**(8), 67–78.

Jeppsson U., Alex J., Batstone D. J., Benedetti L., Comas J., Copp J. B., Corominas Ll., Flores-Alsina X., Gernaey K. V., Nopens I., Pons M.-N., Rodríguez-Roda I., Rosen C., Steyer J.-P., Vanrolleghem P. A., Volcke E. I. P. and Vrecko D. (2013). Benchmark Simulation Models: Quo vadis? *Water Science and Technology*, **68**(1), 1–15.

Lide D. (2004). CRC Handbook of Chemistry and Physics. 85th edn, CRC Press, Boca Raton, FL, USA.

Kleerebezem R. and van Loosdrecht M. C. M. (2006). Waste characterization for implementation in ADM1. *Water Science and Technology*, **54**(4), 167–174.

Maere T., Verrecht B., Moerenhout S., Judd S. and Nopens I. (2011). BSM-MBR: a benchmark simulation model to compare control and operational strategies for membrane bioreactors. *Water Research*, **45**(6), 2181–2190.

Mamdani E. H. and Assilan S. (1975). An experiment in linguistic synthesis with a fuzzy logic controller. *International Man-Machine Studies*, **7**, 1–13.

Martins A. M. P., Heijnen J. J. and van Loosdrecht M. C. M. (2004a). Bulking sludge in biological nutrient removal systems. *Biotechnology and Bioengineering*, **86**(2), 125–135.

Martins A. M. P., Pagilla K., Heijnen J. J. and van Loosdrecht M. C. M. (2004b). Filamentous bulking sludge – A critical review. *Water Research*, **38**(4), 793–817.

Nopens I., Batstone D. J., Copp J. B., Jeppsson U., Volcke E., Alex J. and Vanrolleghem P. A. (2009). An ASM/ADM model interface for dynamic plant-wide simulation. *Water Research*, **43**(7), 1913–1923.

Nopens I., Benedetti L., Jeppsson U., Pons M.-N., Alex J., Copp J. B., Gernaey K. V., Rosen C., Steyer J.-P. and Vanrolleghem P. A. (2010). Benchmark Simulation Model No 2 – Finalisation of plant layout and default control strategy. *Water Science and Technology*, **62**(9), 1967–1974.

Olsson G., Nielsen M., Yuan Z., Lynggaard-Jensen A. and Steyer J. P. (2005). Instrumentation, Control and Automation in Wastewater Systems, IWA Scientific and Technical Report No. 11, IWA Publishing, London, UK.

Olsson G. (2012). ICA and me – A subjective review. *Water Research*, **46**(6), 1585–1624.

Otterpohl R. (1995). Dynamische Simulation zur Unterstützung der Planung und des Betriebes von kommunalen Kläranlagen. PhD thesis, TU Aachen, Germany. Schriftenreihe Gewässerschutz Wasser Abwasser, **151**, 28–36, ISSN 0342-6068.

Plósz B. G., De Clercq J., Nopens I., Benedetti L. and Vanrolleghem P. A. (2011). Shall we upgrade one-dimensional secondary settler models used in WWTP simulators? – An assessment of model structure uncertainty and its propagation. *Water Science and Technology*, **63**(8), 1726–1738.

Pons M. N., Spanjers H. and Jeppsson U. (1999). Towards a benchmark for evaluating control strategies in wastewater treatment plants by simulation. *Computational and Chemical Engineering*, **23**, S403–S406, (ESCAPE9 Supplement).

Reichert P., Borchardt D., Henze M., Rauch W., Shanahan P., Somlyody L. and Vanrolleghem P. A. (2001). River Water Quality Model No. 1 (RWQM1), IWA Scientific and Technical Report No. 12, IWA Publishing, London, UK.

Rieger L., Alex J., Winkler S., Boehler M., Thomann M. and Siegrist H. (2003). Progress in sensor technology – Progress in process control? Part I: Sensor property investigation and classification. *Water Science and Technology*, **47**(2), 103–112.

Rieger L., Alex J., Gujer W. and Siegrist H. (2006). Modelling of aeration systems at wastewater treatment plants. *Water Science and Technology*, **53**(4–5), 439–447.

Rosen C., Röttorp J. and Jeppsson U. (2003). Multivariate on-line monitoring: Challenges and solutions for modern wastewater treatment operation. *Water Science and Technology*, **47**(2), 171–179.

Rosen C., Jeppsson U. and Vanrolleghem P. A. (2004). Towards a common benchmark for long-term process control and monitoring performance evaluation. *Water Science and Technology*, **50**(11), 41–49.

Rosen C., Vrecko D., Gernaey K. V., Pons M. N. and Jeppsson U. (2006). Implementing ADM1 for plant-wide benchmark simulations in Matlab/Simulink. *Water Science and Technology*, **54**(4), 11–19.

Rosen C., Rieger L., Jeppsson U. and Vanrolleghem P. A. (2008). Adding realism to simulated sensors and actuators. *Water Science and Technology*, **57**(3), 337–344.

Rossetti S., Tomei M. C., Nielsen P. H. and Tandoi V. (2005). Microthrix parvicella, a filamentous bacterium causing bulking and foaming in activated sludge systems: a review of current knowledge. *FEMS Microbiology Reviews*, **29**(1), 49–64.

Sharma K. R., Corrie S. and Yuan Z. (2012). Integrated modelling of sewer system and wastewater treatment plant for investigating the impacts of chemical dosing in sewers. *Water Science and Technology*, **65**(8), 1399–1405.

Siegrist H., Krebs P., Bühler R., Purtschert I., Röck C. and Rufer R. (1995). Denitrification in secondary clarifiers. *Water Science and Technology*, **31**(2), 205–214.

Siegrist H., Vogt D., Garcia-Heras J. and Gujer W. (2002). Mathematical model for meso and thermophilic anaerobic sewage sludge digestion. *Environmental Science & Technology*, **36**(5), 1113–1123.

Sin G., Gernaey K. V., Neumann M. B., van Loosdrecht M. C. M. and Gujer W. (2009). Uncertainty analysis in WWTP model applications: A critical discussion using an example from design. *Water Research*, **43**(11), 2894–2906.

Sin G., Gernaey K. V., Neumann M. B., van Loosdrecht M. C. M. and Gujer W. (2011). Global sensitivity analysis in wastewater treatment plant model applications: Prioritizing sources of uncertainty. *Water Research*, **45**(2), 639–651.

Snip L., Flores-Alsina X., Plósz B., Jeppsson U. and Gernaey K. V. (2014), Expanding the BSM platform with occurrence, transport and fate of micropollutants using the ASM-X approach (submitted).

Spanjers H., Vanrolleghem P. A., Nguyen K., Vanhooren H. and Patry G. G. (1998). Towards a simulation-benchmark for evaluating respirometry-based control strategies. *Water Science and Technology*, **37**(12), 219–226.

Spering V., Makinia J. and Rosenwinkel K. H. (2008). Analyzing and improving a mechanistic model for the Microthrix parvicella in activated sludge systems. 1st IWA/WEF Wastewater Treatment Modelling Seminar (WWTmod2008), Mont-Sainte-Anne, Quebec, Canada, June 1–3, 2008.

Stare A., Vrecko D., Hvala N. and Strmcnik S. (2007). Comparison of control strategies for nitrogen removal in an activated sludge process in terms of operating costs: a simulation study. *Water Research*, **41**(9), 2004–2014.

Svärd Å. (2003). Anaerobic Digestion of Urban Organic Waste – Evaluation of Potentials. Licentiate thesis, Department of Water and Environmental Engineering, Lund University, Sweden.

Takács I., Patry G. G. and Nolasco D. (1991). A dynamic model of the clarification thickening process. *Water Research*, **25**(10), 1263–1271.

Vanhooren H. and Nguyen K. (1996). Development of a simulation protocol for evaluation of respirometry-based control strategies. Combined Technical Report of the BIOMATH Department, Ghent University, Belgium and the Department of Civil Engineering, University of Ottawa, Canada.

Vanhooren H., Yuan Z. and Vanrolleghem P. A. (2002). Benchmarking nitrogen removal suspended-carrier biofilm systems using dynamic simulation. *Water Science and Technology*, **46**(1–2), 327–332.

Vanrolleghem P. A., Jeppsson U., Carstensen J., Carlsson B. and Olsson G. (1996). Integration of WWT plant design and operation – A systematic approach using cost functions. *Water Science and Technology*, **34**(3–4), 159–171.

Vanrolleghem P. A. and Gillot S. (2002). Robustness and economic measures as control benchmark performance criteria. *Water Science and Technology*, **45**(4–5), 117–126.

Vanrolleghem P. A., Benedetti L. and Meirlaen J. (2005a). Modelling and real-time control of the integrated urban wastewater system. *Environmental Modelling and Software*, **20**(4), 427–442.

Vanrolleghem P. A., Rosen C., Zaher U., Copp J. B., Benedetti L., Ayesa E. and Jeppsson U. (2005b). Continuity-based interfacing of models for wastewater systems described by Petersen matrices. *Water Science and Technology*, **52**(1–2), 493–500.

Volcke E. I. P., Van Hulle S., Deksissa T., Zaher U. and Vanrolleghem P. A. (2005). Calculation of pH and concentrations of equilibrium components during dynamic simulation by means of a charge balance, Technical Report, BIOMATH, Ghent University, Belgium.

Volcke E. I. P. (2006). Modelling, Analysis and Control of Partial Nitritation in a SHARON Reactor. PhD thesis, Ghent University, Belgium, ISBN 90-5989-108-2.

Volcke E. I. P., Gernaey K. V., Vrecko D., Jeppsson U., van Loosdrecht M. C. M. and Vanrolleghem P. A. (2006a). Plant-wide (BSM2) evaluation of reject water treatment with a SHARON-Anammox process. *Water Science and Technology*, **54**(8), 93–100.

Volcke E. I. P., van Loosdrecht M. C. M. and Vanrolleghem P. A. (2006b). Continuity-based model interfacing for plant-wide simulation: A general approach. *Water Research*, **40**(15), 2817–2828.

Vrecko D., Hvala N. and Kocijan J. (2002). Wastewater treatment benchmark: What can be achieved with simple control? *Water Science and Technology*, **45**(4–5), 127–134.

Wanner J. (1998). Stable foams and sludge bulking: the largest remaining problems (Abriged). *Water and Environment Journal*, **12**(5), 368–374.

Wett B., Eladawy A. and Ogurek M. (2006). Description of nitrogen incorporation and release in ADM1. *Water Science and Technology*, **54**(4), 67–76.

Zaher U., Rongping L., Jeppsson U., Steyer J.-P. and Chen S. (2009). GISCOD: General Integrated Solid waste CO-Digestion model. *Water Research*, **43**(10), 2717–2727.

Appendix A

Model parameters

Tables A.1 to A.12 list the parameters for the various models in the BSMs.

ASM1

Table A.1 Stoichiometric parameter values for ASM1.

Parameter description	Parameter symbol	Value	Units
Autotrophic yield	Y_A	0.24	g $X_{B,A}$ COD formed · (g N oxidised)$^{-1}$
Heterotrophic yield	Y_H	0.67	g $X_{B,H}$ COD formed · (g COD utilised)$^{-1}$
Fraction of biomass to particulate products	f_P	0.08	g X_P COD formed · (g $X_{B,H}$ decayed)$^{-1}$
Fraction nitrogen in biomass	i_{XB}	0.08	g N · (g COD)$^{-1}$ in biomass ($X_{B,A}$ & $X_{B,H}$)
Fraction nitrogen in particulate products	i_{XP}	0.06	g N · (g COD)$^{-1}$ in X_p

Table A.2 Kinetic parameter values for ASM1.

Parameter description	Parameter symbol	Value (@ 15°C)	Temperature coefficient* (a)	Units
Maximum heterotrophic growth rate	μ_{mH}	4.0	3	d^{-1}
Half-saturation (hetero. growth)	K_S	10.0		g COD · m^{-3}
Half-saturation (hetero. oxygen)	K_{OH}	0.2		g O_2 · m^{-3}
Half-saturation (nitrate)	K_{NO}	0.5		g NO_3-N · m^{-3}
Heterotrophic decay rate	b_H	0.3	0.2	d^{-1}

(*Continued*)

Table A.2 Kinetic parameter values for ASM1 (*Continued*).

Parameter description	Parameter symbol	Value (@ 15°C)	Temperature coefficient* (a)	Units
Anoxic growth rate correction factor	η_g	0.8		dimensionless
Anoxic hydrolysis rate correction factor	η_h	0.8		dimensionless
Maximum specific hydrolysis rate	k_h	3.0	2.5	$g\ X_S \cdot (g\ X_{B,H}\ COD \cdot d)^{-1}$
Half-saturation (hydrolysis)	K_X	0.1		$g\ X_S \cdot (g\ X_{B,H}\ COD)^{-1}$
Maximum autotrophic growth rate	μ_{mA}	0.5	0.3	d^{-1}
Half-saturation (auto. growth)	K_{NH}	1.0		$g\ NH_4\text{-}N \cdot m^{-3}$
Autotrophic decay rate	b_A	0.05	0.03	d^{-1}
Half-saturation (auto. oxygen)	K_{OA}	0.4		$g\ O_2 \cdot m^{-3}$
Ammonification rate	k_a	0.05	0.04	$m^3 \cdot (g\ COD \cdot d)^{-1}$

*Kinetic parameters affected by temperature are adjusted according to:

$p_t = p_{15} \cdot \exp((ln(p_{15}/a)/5) \cdot (T - 15))$,

where p_t is the parameter value at temperature T and p_{15} is the parameter value at 15°C.

ADM1

Table A.3 Stoichiometric parameter values for ADM1.

Parameter	BSM value	Units	ADM1 default	Comments
$f_{SI,xc}$	0.1	–		
$f_{XI,xc}$	0.2	–	0.25	
$f_{ch,xc}$	0.2	–		
$f_{pr,xc}$	0.2	–		
$f_{li,xc}$	0.3	–	0.25	*Note*: $1 - f_{ch,xc} - f_{pr,xc} - f_{SI,xc} - f_{li,xc} - f_{XI,xc} = 0$
N_{xc}	0.0376/14	$kmol\ N \cdot kg^{-1}\ COD$	0.002	0.0376/14: to maintain N balance for disintegration
N_I	0.06/14	$kmol\ N \cdot kg^{-1}\ COD$	0.002	*Here*: 6% on weight basis based on ASM1
N_{aa}	0.007	$kmol\ N \cdot kg^{-1}\ COD$		
C_{xc}	0.02786	$kmol\ C \cdot kg^{-1}\ COD$	0.03	to maintain C balance for disintegration
C_{SI}	0.03	$kmol\ C \cdot kg^{-1}\ COD$	0.03	
C_{ch}	0.0313	$kmol\ C \cdot kg^{-1}\ COD$		
C_{pr}	0.03	$kmol\ C \cdot kg^{-1}\ COD$	0.03	

(*Continued*)

Table A.3 Stoichiometric parameter values for ADM1 (*Continued*).

Parameter	BSM value	Units	ADM1 default	Comments
C_{li}	0.022	kmol C · kg^{-1} COD		
C_{XI}	0.03	kmol C · kg^{-1} COD	0.03	
C_{su}	0.0313	kmol C · kg^{-1} COD		
C_{aa}	0.03	kmol C · kg^{-1} COD	0.03	
$f_{fa,li}$	0.95	–		
C_{fa}	0.0217	kmol C · kg^{-1} COD		
$f_{h2,su}$	0.19	–		
$f_{bu,su}$	0.13	–		
$f_{pro,su}$	0.27	–		
$f_{ac,su}$	0.41	–		
N_{bac}	0.08/14	kmol N · kg^{-1} COD	0.00625	*Here*: 8% on weight basis, based on ASM1
C_{bu}	0.025	kmol C · kg^{-1} COD		
C_{pro}	0.0268	kmol C · kg^{-1} COD		
C_{ac}	0.0313	kmol C · kg^{-1} COD		
C_{bac}	0.0313	kmol C · kg^{-1} COD		
Y_{su}	0.1	kmol COD_X · kg^{-1} COD_S		
$f_{h2,aa}$	0.06	–		
$f_{va,aa}$	0.23	–		
$f_{bu,aa}$	0.26	–		
$f_{pro,aa}$	0.05	–		
$f_{ac,aa}$	0.40	–		
C_{va}	0.024	kmol C · kg^{-1} COD		
Y_{aa}	0.08	kmol COD_X · kg^{-1} COD_S		
Y_{fa}	0.06	kmol COD_X · kg^{-1} COD_S		
Y_{c4}	0.06	kmol COD_X · kg^{-1} COD_S		
Y_{pro}	0.04	kmol COD_X · kg^{-1} COD_S		
C_{ch4}	0.0156	kmol C · kg^{-1} COD		
Y_{ac}	0.05	kmol COD_X · kg^{-1} COD_S		
Y_{h2}	0.06	kmol COD_X · kg^{-1} COD_S		

Table A.4 Kinetic parameter values for ADM1.

Parameter	Value	Units
k_{dis}	0.5	d^{-1}
$k_{hyd,ch}$	10	d^{-1}
$k_{hyd,pr}$	10	d^{-1}
$k_{hyd,li}$	10	d^{-1}
$K_{S,IN}$	10^{-4}	kmol $N \cdot m^{-3}$
$k_{m,su}$	30	d^{-1}
$K_{S,su}$	0.5	kg $COD \cdot m^{-3}$
$pH_{UL,aa}$	5.5	–
$pH_{LL,aa}$	4	–
$k_{m,aa}$	50	d^{-1}
$K_{S,aa}$	0.3	kg $COD \cdot m^{-3}$
$k_{m,fa}$	6	d^{-1}
$K_{S,fa}$	0.4	kg $COD \cdot m^{-3}$
$K_{I,h2,fa}$	$5 \cdot 10^{-6}$	kg $COD \cdot m^{-3}$
$k_{m,c4}$	20	d^{-1}
$K_{S,c4}$	0.2	kg $COD \cdot m^{-3}$
$K_{I,h2,c4}$	10^{-5}	kg $COD \cdot m^{-3}$
$k_{m,pro}$	13	d^{-1}
$K_{S,pro}$	0.1	kg $COD \cdot m^{-3}$
$K_{I,h2,pro}$	$3.5 \cdot 10^{-6}$	kg $COD \cdot m^{-3}$
$k_{m,ac}$	8	d^{-1}
$K_{S,ac}$	0.15	kg $COD \cdot m^{-3}$
$K_{I,NH3}$	0.0018	kmol $NH_4\text{-}N \cdot m^{-3}$
$pH_{UL,ac}$	7	–
$pH_{LL,ac}$	6	–
$k_{m,h2}$	35	d^{-1}
$K_{S,h2}$	$7 \cdot 10^{-6}$	kg $COD \cdot m^{-3}$
$pH_{UL,h2}$	6	–
$pH_{LL,h2}$	5	–
$k_{dec,Xsu}$	0.02	d^{-1}
$k_{dec,Xaa}$	0.02	d^{-1}
$k_{dec,Xfa}$	0.02	d^{-1}
$k_{dec,Xc4}$	0.02	d^{-1}
$k_{dec,Xpro}$	0.02	d^{-1}
$k_{dec,Xac}$	0.02	d^{-1}
$k_{dec,Xh2}$	0.02	d^{-1}

Table A.5 Physico-chemical parameter values for the ADM1.

Parameter	Constant	Temperature coefficients		Units	Comments
		a	b		
R	0.083145			$bar \cdot M^{-1} \cdot K^{-1}$	ADM1 default value = 0.08314
T_{base}	298.15			K	ADM1 default value = 298
T_{ad}	308.15			K	=35°C
K_w		10^{-14}	55 900	M	$\approx 2.08 \cdot 10^{-14}$ at 35°C
$K_{a,va}$	$10^{-4.86}$			M	$\approx 1.38 \cdot 10^{-5}$
$K_{a,bu}$	$10^{-4.82}$			M	$\approx 1.5 \cdot 10^{-5}$
$K_{a,pro}$	$10^{-4.88}$			M	$\approx 1.32 \cdot 10^{-5}$
$K_{a,ac}$	$10^{-4.76}$			M	$\approx 1.74 \cdot 10^{-5}$
$K_{a,co2}$		$10^{-6.35}$	7 646	M	$\approx 4.94 \cdot 10^{-7}$ at 35°C
$K_{a,IN}$		$10^{-9.25}$	51 965	M	$\approx 1.11 \cdot 10^{-9}$ at 35°C
$k_{A,Bva}$	10^{10}			$M^{-1} \cdot d^{-1}$	Set to be at least three orders of magnitude higher than the fastest time constant of the system ADM1 suggested default value = 1.10^8
$k_{A,Bbu}$	10^{10}			$M^{-1} \cdot d^{-1}$	
$k_{A,Bpro}$	10^{10}			$M^{-1} \cdot d^{-1}$	
$k_{A,Bac}$	10^{10}			$M^{-1} \cdot d^{-1}$	
$k_{A,Bco2}$	10^{10}			$M^{-1} \cdot d^{-1}$	
$k_{A,BIN}$	10^{10}			$M^{-1} \cdot d^{-1}$	
P_{atm}	1.013			bar	
$p_{gas,h2o}$		0.0313	$529\,000 \cdot R$	bar	≈ 0.0557 at 35°C
k_p	$5 \cdot 10^4$			$m^3 \cdot d^{-1} \cdot bar^{-1}$	Explicit for BSM2 AD conditions. Must be recalibrated for other cases.
K_La	200			d^{-1}	
$K_{H,co2}$		0.035	−19 410	$M_{liq} \cdot bar^{-1}$	≈ 0.0271 at 35°C
$K_{H,ch4}$		0.0014	−14 240	$M_{liq} \cdot bar^{-1}$	≈ 0.00116 at 35°C
$K_{H,h2}$		$7.8 \cdot 10^{-4}$	−4 180	$M_{liq} \cdot bar^{-1}$	$\approx 7.38 \cdot 10^{-4}$ at 35°C

*Parameters affected by temperature are calculated according to:

$p_t = a \cdot \exp(b/100 \cdot R \cdot (1/T_{base} - 1/T_{ad}))$,

where p_t is the parameter value at temperature T_{ad}. Subscript liq stands for 'liquid phase'.

ASM1/ADM INTERFACE

Table A.6 ASM1/ADM interface model parameters.

Parameter description	Symbol	Value	Units	Interface step
Nitrogen content of amino acids	N_{aa}	0.007	mol N·g COD^{-1}	2
Lipid fraction of non-nitrogenous X_S	f_{rlixs}	0.7	–	3
Anaerobic degradable fraction of biomass	f_{rxi}	0.68	–	4
Biomass nitrogen content	i_{nx}	0.08	mol N·g COD^{-1}	4
Lipid fraction of non-nitrogenous biomass	f_{rlixb}	0.4	–	4
Nitrogen content of soluble inerts in ASM1	n_{si}	0	mol N·g COD^{-1}	5
Nitrogen content of soluble inerts in ADM1	n_{si_adm}	0.06	mol N·g COD^{-1}	5

SOLIDS/LIQUID SEPARATION PROCESSES

Table A.7 Primary clarifier model parameters used to achieve 50% TSS removal and 3% TSS concentration in the underflow.

Parameter description	Symbol	Value	Units
Underflow proportion of influent flow	f_{Qu}	0.007	–
Correction factor removal efficiency (tuning parameter)	f_{corr}	0.65	–
Ratio of primary sludge flow rate to the influent flow rate	f_{pu}	0.01	–
Ratio of CODpart to CODtot (mean value)	f_X	0.75	–
Soluble separation factor	$f_{SX,s}$	0	–
Particulate separation factor	$f_{SX,x}$	1.0	–
Particulate fraction of X_S	f_{XS}	0.5	–
Smoothing time constant for average flow calculation (3 h)	T_m	3/24	d

Table A.8 Secondary clarifier model parameters.

Parameter description	Parameter symbol	Value	Units
Maximum settling velocity	v'_o	250	m·d^{-1}
Maximum Vesilind settling velocity	v_o	474	m·d^{-1}
Hindered zone settling parameter	r_h	0.000576	m^3·(g SS)$^{-1}$
Flocculant zone settling parameter	r_p	0.00286	m^3·(g SS)$^{-1}$
Non-settleable fraction	f_{ns}	0.00228	dimensionless

SENSORS

Table A.9 Typical sensor characteristics within the proposed classification scheme.

Measured variable	Sensor class	t_d [min]	t_l [min]
MLSS ($g \cdot m^{-3}$)	A	0	0
Turbidity (FNU or g TSS $\cdot m^{-3}$)	A		
S_{NH4} (ion sensitive)	A		
S_{NOx} (ion sensitive)	A		
S_{NOx} (UV)	A		
C_{COD}, S_{COD} (UV/Vis)	A		
Flow rate ($m^3 \cdot d^{-1}$)	A		
Water level (m)	A		
Temperature (°C)	A		
pH	A		
S_O ($g\ O_2 \cdot m^{-3}$)	A		
Sludge blanket height (m)	A		
S_{NH4} (gas sensitive + normal filtration)	B_0	10	0
S_{NOx} (UV + normal filtration)	B_0		
S_{NH4} (photometric + normal filtration)	B_1	10	5
S_{NO3} (photometric + normal filtration)	B_1		
S_{NO2} (photometric + normal filtration)	B_1		
S_{PO4} (photometric + normal filtration)	B_1		
S_{NH4} (gas sensitive + slow filtration or sedimentation)	C_0	20	0
S_{NOx} (UV + slow filtration or sedimentation)	C_0		
S_{NH4} (photometric + slow filtration or sedimentation)	C_1	20	5
S_{NO3} (photometric + slow filtration or sedimentation)	C_1		
S_{NO2} (photometric + slow filtration or sedimentation)	C_1		
S_{PO4} (photometric + slow filtration or sedimentation)	C_1		
C_{COD} (thermal chemical oxidation + photometric)	D	30	30
TOC (thermal oxidation + IR detector)	D		
C_N (thermal oxidation + IR detector or chemoluminescence detector)	D		
C_P (thermal chemical oxidation + photometric)	D		
Respirometer	D		
Titration biosensor (alkalinity, VFA)	D		

Table A.10 Recommended sensor parameter values for BSM1 and BSM2.

Measured variable	Class	Measurement range	Measurement noise (δ)
Flow rate ($m^3 \cdot d^{-1}$) high range	A	0–100 000	2 500
Water level (m)	A	0–5	0.125
Temperature (°C)	A	5–25	0.5
pH	A	5–9	0.1
S_O (g $O_2 \cdot m^{-3}$)	A	0–10	0.25
Sludge blanket level (m)	A	0–5	0.125
S_{NO} (g $N \cdot m^{-3}$)	B_0	0–20	0.5
S_{NH} (g $N \cdot m^{-3}$) low range	B_0	0–20	0.5
S_{NH} (g $N \cdot m^{-3}$) high range	B_0	0–50	1.25
S_{ALK} (mol $HCO_3 \cdot m^{-3}$)	B_0	0–20	0.5
Mixed-liquor suspended solids (g $\cdot m^{-3}$)	A	0–10 000	250
Effluent total suspended solids (g $\cdot m^{-3}$)	A	0–200	5
COD_{tot} (g COD $\cdot m^{-3}$)	D	0–1 000	25
OUR (g $O_2 \cdot m^{-3} \cdot d^{-1}$)	D	0–2 000	50

Table A.11 Parameters for response time modelling of continuously operating sensors.

Sensor class	t_r (min)	n	τ (min)	$R_{td/tr}$
A	1	2	0.257	0.133
B_0	10	8	0.849	0.392
C_0	20	8	1.699	0.392

AVAILABLE CONTROL HANDLES

Table A.12 Available control handles and their limitations for BSM1 (all control handles have a minimum value of 0).

Control handle	Maximum value	Units	Comments
Q_{int}	92 230	$m^3 \cdot d^{-1}$	Max = 500% of $Q_{0,stab}$
Q_r	36 892	$m^3 \cdot d^{-1}$	Max = 200% of $Q_{0,stab}$
Q_w	1 844.6	$m^3 \cdot d^{-1}$	Max = 10% of $Q_{0,stab}$
K_La_1	360	d^{-1}	Reactor 1
K_La_2	360	d^{-1}	Reactor 2

(*Continued*)

Table A.12 Available control handles and their limitations for BSM1 (all control handles have a minimum value of 0) (*Continued*).

Control handle	Maximum value	Units	Comments
K_La_3	360	d^{-1}	Reactor 3
K_La_4	360	d^{-1}	Reactor 4
K_La_5	360	d^{-1}	Reactor 5
Q_{EC1}	5	$m^3 \cdot d^{-1}$	Reactor 1
Q_{EC2}	5	$m^3 \cdot d^{-1}$	Reactor 2
Q_{EC3}	5	$m^3 \cdot d^{-1}$	Reactor 3
Q_{EC4}	5	$m^3 \cdot d^{-1}$	Reactor 4
Q_{EC5}	5	$m^3 \cdot d^{-1}$	Reactor 5
f_{Qi1}, f_{Qi2}, f_{Qi3}, f_{Qi4}, f_{Qi5}	1		influent flow rate distribution. *Note*: the sum of all five must always equal one
f_{Qint1}, f_{Qint2}, f_{Qint3}, f_{Qint4}, f_{Qint5}	1		internal recirculation flow rate distribution. *Note*: the sum of all five must always equal one
f_{Qr1}, f_{Qr2}, f_{Qr3}, f_{Qr4}, f_{Qr5}	1		sludge return flow rate distribution. *Note*: the sum of all five must always equal one

Appendix B

Simulation output

I. Nopens, W. De Keyser, L. Corominas, L. Benedetti, M.-N. Pons, J. Alex, J. B. Copp, J. Dudley, C. Rosen, P. A. Vanrolleghem and U. Jeppsson

This Appendix lists the simulation output generated by the benchmark systems. Note that only steady state results in open-loop configuration are given here and that results from BSM1_LT are not provided. Ring-testing of BSM1_LT is on-going but all results will soon be made available in an updated version of Technical Report No. 14. In that technical report all results are listed for all platforms for all different conditions (steady state, dynamic open-loop, dynamic ideal and non-ideal closed-loop cases).

STEADY STATE RESULTS

BSM1

The results of the steady state ring-test are summarised in Tables B.1 to B.13. The tables list some of the values obtained by each software platform. The results indicate that all of the implementations yield essentially identical results. The very small deviations are attributed to rounding and numerical errors, but this was not definitively verified.

Table B.1 Effluent concentrations for BSM1 steady state.

Variable	Unit	WEST	FORTRAN	MATLAB	SIMBA	GPS-X	Average
Q_e	$m^3 \cdot d^{-1}$	18 061	18 061	18 061	18 061	18 061	18 061
S_I	$g\ COD \cdot m^{-3}$	30	30	30	30	30	30
S_S	$g\ COD \cdot m^{-3}$	0.8895	0.8876	0.8895	0.8895	0.8895	0.8891
X_I	$g\ COD \cdot m^{-3}$	4.392	4.392	4.392	4.392	4.392	4.392
X_S	$g\ COD \cdot m^{-3}$	0.188	0.188	0.188	0.188	0.188	0.188
$X_{B,H}$	$g\ COD \cdot m^{-3}$	9.782	9.781	9.782	9.782	9.782	9.782

(*Continued*)

Table B.1 Effluent concentrations for BSM1 steady state (*Continued*).

Variable	Unit	WEST	FORTRAN	MATLAB	SIMBA	GPS-X	Average
$X_{B,A}$	g COD·m^{-3}	0.573	0.573	0.573	0.573	0.573	0.573
X_P	g COD·m^{-3}	1.728	1.728	1.728	1.728	1.728	1.728
S_O	g O_2·m^{-3}	0.491	0.491	0.491	0.491	0.491	0.491
S_{NO}	g N·m^{-3}	10.415	10.408	10.415	10.415	10.415	10.414
S_{NH}*	g N·m^{-3}	1.733	1.744	1.733	1.733	1.733	1.735
S_{ND}	g N·m^{-3}	0.688	0.689	0.688	0.688	0.688	0.689
X_{ND}	g N·m^{-3}	0.013	0.013	0.013	0.013	0.013	0.013
S_{ALK}	mol HCO_3·m^{-3}	4.126	4.127	4.126	4.126	4.126	4.126
TSS*	g SS·m^{-3}	12.497	12.497	12.497	12.497	12.497	12.497
S_{NKj}	g N·m^{-3}	3.631	3.629	3.631	3.631	3.631	3.630
N_{tot}*	g N·m^{-3}	14.046	14.036	14.046	14.046	14.046	14.044
COD_{tot}*	g COD·m^{-3}	47.552	47.550	47.552	47.552	47.552	47.552
BOD_5*	g O_2·m^{-3}	2.651	2.650	2.651	2.651	2.651	2.651

*limits are: S_{NH} = 4 g N·m^{-3}; TSS = 30 g·m^{-3}; N_{tot} = 18 g N·m^{-3}; COD_{tot} = 100 g COD·m^{-3}; BOD_5 = 10 g O_2·m^{-3}.

Table B.2 Concentrations in reactor 1 for BSM1 steady state.

Variable	Unit	WEST	FORTRAN	MATLAB	SIMBA	GPS-X	Average
S_I	g COD·m^{-3}	30	30	30	30	30	30
S_S	g COD·m^{-3}	2.808	2.802	2.808	2.808	2.808	2.807
X_I	g COD·m^{-3}	1 149.125	1 149.140	1 149.125	1 149.125	1 149.119	1 149.127
X_S	g COD·m^{-3}	82.135	82.137	82.135	82.135	82.135	82.135
$X_{B,H}$	g COD·m^{-3}	2 551.766	2 551.760	2 551.766	2 551.766	2 551.764	2 551.764
$X_{B,A}$	g COD·m^{-3}	148.389	148.409	148.389	148.389	148.391	148.394
X_P	g COD·m^{-3}	448.852	448.860	448.852	448.852	448.849	448.853
S_O	g O_2·m^{-3}	0.004	0.004	0.004	0.004	0.004	0.004
S_{NO}	g N·m^{-3}	5.370	5.363	5.370	5.370	5.370	5.369
S_{NH}	g N·m^{-3}	7.918	7.925	7.918	7.918	7.918	7.919
S_{ND}	g N·m^{-3}	1.217	1.218	1.217	1.217	1.217	1.217
X_{ND}	g N·m^{-3}	5.285	5.285	5.285	5.285	5.285	5.285
S_{ALK}	mol HCO_3·m^{-3}	4.928	4.929	4.928	4.928	4.928	4.928
TSS	g SS·m^{-3}	3 285.200	3 285.230	3 285.200	3 285.200	3 285.190	3 285.205

Table B.3 Concentrations in reactor 2 for BSM1 steady state.

Variable	Unit	WEST	FORTRAN	MATLAB	SIMBA	GPS-X	Average
S_I	g COD·m^{-3}	30	30	30	30	30	30
S_S	g COD·m^{-3}	1.459	1.455	1.459	1.459	1.459	1.458
X_I	g COD·m^{-3}	1 149.125	1 149.140	1 149.125	1 149.125	1 149.119	1 149.127

(*Continued*)

Table B.3 Concentrations in reactor 2 for BSM1 steady state (*Continued*).

Variable	Unit	WEST	FORTRAN	MATLAB	SIMBA	GPS-X	Average
X_S	g COD·m^{-3}	76.386	76.390	76.386	76.386	76.386	76.387
$X_{B,H}$	g COD·m^{-3}	2 553.385	2 553.380	2 553.385	2 553.385	2 553.383	2 553.384
$X_{B,A}$	g COD·m^{-3}	148.309	148.329	148.309	148.309	148.310	148.313
X_P	g COD·m^{-3}	449.523	449.531	449.523	449.523	449.520	449.524
S_O	g O_2·m^{-3}	0.000	0.000	0.000	0.000	0.000	0.000
S_{NO}	g N·m^{-3}	3.662	3.656	3.662	3.662	3.662	3.661
S_{NH}	g N·m^{-3}	8.344	8.352	8.344	8.344	8.344	8.346
S_{ND}	g N·m^{-3}	0.882	0.883	0.882	0.882	0.882	0.882
X_{ND}	g N·m^{-3}	5.029	5.029	5.029	5.029	5.029	5.029
S_{ALK}	mol HCO_3·m^{-3}	5.080	5.081	5.080	5.080	5.080	5.080
TSS	g SS·m^{-3}	3 282.546	3 282.578	3 282.546	3 282.546	3 282.539	3 282.551

Table B.4 Concentrations in reactor 3 for BSM1 steady state.

Variable	Unit	WEST	FORTRAN	MATLAB	SIMBA	GPS-X	Average
S_I	g COD·m^{-3}	30	30	30	30	30	30
S_S	g COD·m^{-3}	1.150	1.147	1.150	1.150	1.150	1.149
X_I	g COD·m^{-3}	1 149.125	1 149.140	1 149.125	1 149.125	1 149.119	1 149.127
X_S	g COD·m^{-3}	64.855	64.857	64.855	64.855	64.855	64.855
$X_{B,H}$	g COD·m^{-3}	2 557.131	2 557.130	2 557.131	2 557.131	2 557.129	2 557.131
$X_{B,A}$	g COD·m^{-3}	148.941	148.961	148.941	148.941	148.943	148.945
X_P	g COD·m^{-3}	450.418	450.426	450.418	450.418	450.416	450.419
S_O	g O_2·m^{-3}	1.718	1.722	1.718	1.718	1.718	1.719
S_{NO}	g N·m^{-3}	6.541	6.533	6.541	6.541	6.541	6.539
S_{NH}	g N·m^{-3}	5.548	5.558	5.548	5.548	5.548	5.550
S_{ND}	g N·m^{-3}	0.829	0.830	0.829	0.829	0.829	0.829
X_{ND}	g N·m^{-3}	4.392	4.393	4.392	4.392	4.392	4.392
S_{ALK}	mol HCO_3·m^{-3}	4.675	4.676	4.675	4.675	4.675	4.675
TSS	g SS·m^{-3}	3 277.853	3 277.886	3 277.853	3 277.853	3 277.846	3 277.858

Table B.5 Concentrations in reactor 4 for BSM1 steady state.

Variable	Unit	WEST	FORTRAN	MATLAB	SIMBA	GPS-X	Average
S_I	g COD·m^{-3}	30	30	30	30	30	30
S_S	g COD·m^{-3}	0.995	0.993	0.995	0.995	0.995	0.995
X_I	g COD·m^{-3}	1 149.125	1 149.140	1 149.125	1 149.125	1 149.119	1 149.127
X_S	g COD·m^{-3}	55.694	55.695	55.694	55.694	55.694	55.694
$X_{B,H}$	g COD·m^{-3}	2 559.183	2 559.180	2 559.183	2 559.183	2 559.181	2 559.182

(*Continued*)

Table B.5 Concentrations in reactor 4 for BSM1 steady state (*Continued*).

Variable	Unit	WEST	FORTRAN	MATLAB	SIMBA	GPS-X	Average
$X_{B,A}$	g COD·m^{-3}	149.527	149.547	149.527	149.527	149.528	149.531
X_P	g COD·m^{-3}	451.315	451.323	451.315	451.315	451.312	451.316
S_O	g O_2·m^{-3}	2.429	2.432	2.429	2.429	2.429	2.429
S_{NO}	g N·m^{-3}	9.299	9.290	9.299	9.299	9.299	9.297
S_{NH}	g N·m^{-3}	2.967	2.979	2.967	2.967	2.967	2.970
S_{ND}	g N·m^{-3}	0.767	0.768	0.767	0.767	0.767	0.767
X_{ND}	g N·m^{-3}	3.879	3.879	3.879	3.879	3.879	3.879
S_{ALK}	mol HCO_3·m^{-3}	4.293	4.295	4.294	4.293	4.293	4.294
TSS	g SS·m^{-3}	3 273.633	3 273.664	3 273.633	3 273.633	3 273.625	3 273.637

Table B.6 Concentrations in reactor 5 for BSM1 steady state.

Variable	Unit	WEST	FORTRAN	MATLAB	SIMBA	GPS-X	Average
S_I	g COD·m^{-3}	30	30	30	30	30	30
S_S	g COD·m^{-3}	0.889	0.888	0.889	0.889	0.889	0.889
X_I	g COD·m^{-3}	1 149.125	1 149.140	1 149.125	1 149.125	1 149.119	1 149.127
X_S	g COD·m^{-3}	49.306	49.307	49.306	49.306	49.305	49.306
$X_{B,H}$	g COD·m^{-3}	2 559.344	2 559.340	2 559.344	2 559.344	2 559.342	2 559.343
$X_{B,A}$	g COD·m^{-3}	149.797	149.817	149.797	149.797	149.798	149.801
X_P	g COD·m^{-3}	452.211	452.219	452.211	452.211	452.209	452.212
S_O	g O_2·m^{-3}	0.491	0.491	0.491	0.491	0.491	0.491
S_{NO}	g N·m^{-3}	10.415	10.408	10.415	10.415	10.415	10.414
S_{NH}	g N·m^{-3}	1.733	1.744	1.733	1.733	1.733	1.735
S_{ND}	g N·m^{-3}	0.688	0.689	0.688	0.688	0.688	0.689
X_{ND}	g N·m^{-3}	3.527	3.527	3.527	3.527	3.527	3.527
S_{ALK}	mol HCO_3·m^{-3}	4.126	4.127	4.126	4.126	4.126	4.126
TSS	g SS·m^{-3}	3 269.837	3 269.867	3 269.837	3 269.837	3 269.829	3 269.841

Table B.7 Concentrations in settler underflow for BSM1 steady state.

Variable	Unit	WEST	FORTRAN	MATLAB	SIMBA	GPS-X	Average
S_I	g COD·m^{-3}	30	30	30	30	30	30
S_S	g COD·m^{-3}	0.889	0.888	0.889	0.889	0.889	0.889
X_I	g COD·m^{-3}	2 247.050	2 247.090	2 247.050	2 247.050	2 247.038	2 247.056
X_S	g COD·m^{-3}	96.414	96.416	96.414	96.414	96.414	96.415
$X_{B,H}$	g COD·m^{-3}	5 004.654	5 004.650	5 004.654	5 004.654	5 004.650	5 004.652
$X_{B,A}$	g COD·m^{-3}	292.920	292.959	292.920	292.920	292.922	292.928
X_P	g COD·m^{-3}	884.274	884.289	884.274	884.274	884.269	884.276

(*Continued*)

Table B.7 Concentrations in settler underflow for BSM1 steady state (*Continued*).

Variable	Unit	WEST	FORTRAN	MATLAB	SIMBA	GPS-X	Average
S_O	g $O_2 \cdot m^{-3}$	0.491	0.491	0.491	0.491	0.491	0.491
S_{NO}	g $N \cdot m^{-3}$	10.415	10.408	10.415	10.415	10.415	10.414
S_{NH}	g $N \cdot m^{-3}$	1.733	1.744	1.733	1.733	1.733	1.735
S_{ND}	g $N \cdot m^{-3}$	0.688	0.689	0.688	0.688	0.688	0.689
X_{ND}	g $N \cdot m^{-3}$	6.897	6.897	6.897	6.897	6.897	6.897
S_{ALK}	mol $HCO_3 \cdot m^{-3}$	4.126	4.127	4.126	4.126	4.126	4.126
TSS	g $SS \cdot m^{-3}$	6 393.984	6 394.050	6 393.984	6 393.984	6 393.969	6 393.995

Table B.8 Concentrations in secondary settling tank for BSM1 steady state (number stands for layer, where 1 indicates the top layer and 10 the bottom layer).

Variable	Unit	WEST	FORTRAN	MATLAB	SIMBA	GPS-X	Average
TSS1	g $SS \cdot m^{-3}$	12.497	12.497	12.497	12.497	12.497	12.497
TSS2	g $SS \cdot m^{-3}$	18.113	18.113	18.113	18.113	18.113	18.113
TSS3	g $SS \cdot m^{-3}$	29.540	29.540	29.540	29.540	29.540	29.540
TSS4	g $SS \cdot m^{-3}$	68.978	68.978	68.978	68.978	68.978	68.978
TSS5	g $SS \cdot m^{-3}$	356.075	356.077	356.075	356.075	356.074	356.075
TSS6	g $SS \cdot m^{-3}$	356.075	356.077	356.075	356.075	356.074	356.075
TSS7	g $SS \cdot m^{-3}$	356.075	356.077	356.075	356.075	356.074	356.075
TSS8	g $SS \cdot m^{-3}$	356.075	356.077	356.075	356.075	356.074	356.075
TSS9	g $SS \cdot m^{-3}$	356.075	356.077	356.075	356.075	356.074	356.075
TSS10	g $SS \cdot m^{-3}$	6 393.984	6 394.050	6 393.984	6 393.984	6 393.969	6 393.995

Table B.9 Output quality variables for BSM1 steady state. IQI orig and EQI orig represent values based on the original BSM1 evaluation criteria calculation method whereas IQI upd and EQI upd represent the currently used method for those calculations (now identical for BSM1, BSM1_LT and BSM2).

Variable	Unit	WEST	FORTRAN	MATLAB	SIMBA	GPS-X	Average
IQI orig	kg $PU \cdot d^{-1}$	42 043.86		42 043.86			42 043.86
IQI upd	kg $PU \cdot d^{-1}$	52 083.21	52 100.00	52 083.21	52 083.21	52 083.21	52 086.57
EQI orig	kg $PU \cdot d^{-1}$	6 479.65		6 479.65			6 479.65
EQI upd	kg $PU \cdot d^{-1}$	5 254.28	5 260.00	5 254.28	5 254.28	5 254.25	5 255.42
Sludge for disposal	kg $SS \cdot d^{-1}$	2 461.68		2 461.68	2 461.68	2 461.68	2 461.68

(*Continued*)

Table B.9 Output quality variables for BSM1 steady state. IQI orig and EQI orig represent values based on the original BSM1 evaluation criteria calculation method whereas IQI upd and EQI upd represent the currently used method for those calculations (now identical for BSM1, BSM1_LT and BSM2) (*Continued*).

Variable	Unit	WEST	FORTRAN	MATLAB	SIMBA	GPS-X	Average
Sludge into effluent	kg SS · d^{-1}	225.71		225.71	225.71	225.71	225.71
Total sludge	kg SS · d^{-1}	2 687.39	2 687.00	2 687.39	2 687.39	2 687.39	2 687.31

IQI: Influent Quality Index, EQI: Effluent Quality Index.

Table B.10 Energy related variables for BSM1 steady state. Aeration orig and Pumping orig represent values based on the original BSM1 evaluation criteria calculation method whereas Aeration upd and Pumping upd represent the currently used method for those calculations (now identical for BSM1, BSM1_LT and BSM2).

Variable	Unit	WEST	FORTRAN	MATLAB	SIMBA	GPS-X	Average
Aeration orig	kWh · d^{-1}	6 476.11		6 476.11			6 476.11
Aeration upd	kWh · d^{-1}	3 341.39	3 341.0	3 341.39	3 341.39	3 341.39	3 341.31
Pumping orig	kWh · d^{-1}	2 966.76		2 966.76			2 966.76
Pumping upd	kWh · d^{-1}	3 88.17	388.2	388.17	388.17	388.17	388.18
Carbon add	kg COD · d^{-1}	0.00	0.00	0.00	0.00	0.00	0.00
Mixing	kWh · d^{-1}	240.00	240.0	240.00	240.00	240.00	240.00

Table B.11 Operational cost index calculation for BSM1 steady state. AE orig, PE orig and OCI orig represent values based on the original BSM1 evaluation criteria calculation method whereas AE upd, PE upd and OCI upd represent the currently used method for those calculations (now identical for BSM1, BSM1_LT and BSM2). Note that weight factors are used to define the indices (therefore no unit): *SP* = Total sludge production per day * 5, *EC* = External carbon addition per day * 3. For all other indices the weight factor is 1. OCI represents the sum of the individual indices.

Variable	Unit	WEST	FORTRAN	MATLAB	SIMBA	GPS-X	Average
SP	–	12 308.420		12 308.42		12 308.39	12 308.41
AE orig	–	6 476.112		6 476.11			6 476.11
AE upd	–	3 341.387	3 341.0	3 341.39	3 341.39	3 341.39	3 341.31
PE orig	–	2 966.760		2 966.76			2 966.76
PE upd	–	388.170	388.20	388.17	388.17	388.17	388.18
EC	–	0.000	0.00	0.00	0.00	0.00	0.00
ME	–	240.000	240.00	240.00	240.00	240.00	240.00
OCI orig	–	21 991.292		21 991.29			21 991.29
OCI upd	–	16 277.977	16 280.00	16 277.98	16 277.98	16 277.95	16 278.38

SP: Sludge Production index, *AE*: Aeration Energy index, *PE*: Pumping Energy index, *EC*: External Carbon addition index, *ME*: Mixing Energy index, OCI: Operational Cost Index.

Table B.12 Effluent violations for BSM1 steady state.

Variable	Unit	WEST	FORTRAN	MATLAB	SIMBA	GPS-X	Average
$S_{NH,e95}$	$g\ N \cdot m^{-3}$	1.733	1.740	1.733	1.733	1.733	1.735
$N_{tot,e95}$	$g\ N \cdot m^{-3}$	14.046	14.000	14.046	14.046	14.046	14.037
TSS_{e95}	$g\ SS \cdot m^{-3}$	12.497	12.500	12.497	12.497	12.497	12.498
N_{tot} viol.*	d	0.000	0.000	0.000	0.000	0.000	0.000
	%	0.000	0.000	0.000	0.000	0.000	0.000
	–	0.000	0.000	0.000	0.000	0.000	0.000
COD_{tot}	d	0.000	0.000	0.000	0.000	0.000	0.000
	%	0.000	0.000	0.000	0.000	0.000	0.000
	–	0.000	0.000	0.000	0.000	0.000	0.000
S_{NH}	d	0.000	0.000	0.000	0.000	0.000	0.000
	%	0.000	0.000	0.000	0.000	0.000	0.000
	–	0.000	0.000	0.000	0.000	0.000	0.000
TSS	d	0.000	0.000	0.000	0.000	0.000	0.000
	%	0.000	0.000	0.000	0.000	0.000	0.000
	–	0.000	0.000	0.000	0.000	0.000	0.000
BOD_5	d	0.000	0.000	0.000	0.000	0.000	0.000
	%	0.000	0.000	0.000	0.000	0.000	0.000
	–	0.000	0.000	0.000	0.000	0.000	0.000

*limits are: $S_{NH} = 4\ g\ N \cdot m^{-3}$; $TSS = 30\ g \cdot m^{-3}$; $N_{tot} = 18\ g\ N \cdot m^{-3}$; $COD_{tot} = 100\ g\ COD \cdot m^{-3}$; $BOD_5 = 10\ g\ O_2 \cdot m^{-3}$; violations are expressed as duration in days, % of time of total evaluation period, number of exceedances of limit.

Table B.13 Risk results for microbiology-related settling problems for BSM1 steady state (only available in Fortran and Matlab BSM1 implementations).

Individual risk	FORTRAN	MATLAB	Average
Bulking due to low DO values (average raw risk)	0.245	0.246	0.245
Bulking due to low DO values (%time in severe raw risk, >0.8)	0.000	0.000	0.000
Bulking due to low organic loading (average raw risk)	0.654	0.654	0.654
Bulking due to low organic loading (%time in severe raw risk, >0.8)	0.000	0.000	0.000
Bulking due to nutrient deficiency (average raw risk)	0.000	0.000	0.000
Bulking due to nutrient deficiency (%time in severe raw risk, >0.8)	0.000	0.000	0.000
Foaming due to low F/M ratio (average raw risk)	0.583	0.583	0.583
Foaming due to low F/M ratio (%time in severe raw risk, 0.8)	0.000	0.000	0.000
Foaming due to high RBOM* fraction (average raw risk)	0.000	0.000	0.000
Foaming due to high RBOM* fraction (%time in severe raw risk, >0.8)	0.000	0.000	0.000
Rising sludge (average raw risk)	1.000	1.000	1.000
Rising sludge (%time in severe raw risk, >0.8)	100.000	100.000	100.000

(*Continued*)

Table B.13 Risk results for microbiology-related settling problems for BSM1 steady state (only available in Fortran and Matlab BSM1 implementations) (*Continued*).

Individual risk	FORTRAN	MATLAB	Average
Integrated risks			
Bulking (average risk)	0.654	0.654	0.654
Bulking (%time in severe risk, >0.8)	0.000	0.000	0.000
Foaming (average risk)	0.583	0.583	0.583
Foaming (%time in severe risk, >0.8)	0.000	0.000	0.000
Rising (average risk)	1.000	1.000	1.000
Rising (%time in severe risk, >0.8)	100.000	100.000	100.000
Overall Risk (average risk)	1.000	1.000	1.000
Overall Risk (%time in severe risk, >0.8)	100.000	100.000	100.000

**Readily biodegradable organic matter*

BSM2

As BSM2 consists of a number of complicated subsystems, ring-tests were first performed for the separate units including the primary clarifier, thickener, ASM to ADM interface, anaerobic digester, ADM to ASM interface and the dewatering unit. Finally, a ring test was performed on the full BSM2 system in steady state, dynamic open-loop, dynamic ideal and non-ideal closed loop. A sampling of the results for steady state conditions generated in these ring-tests is presented here, Tables B.14 to B.37. For the complete results see Technical Report No. 14.

Primary clarifier

The primary clarifier implementation was tested for two steady state influent cases as shown in Table B.14. Test results are identical.

Table B.14 Influent composition of the two ring-test cases for the primary clarifier.

Variable	Unit	Case 1	Case 2
S_I	$g\ COD \cdot m^{-3}$	30	30
S_S	$g\ COD \cdot m^{-3}$	69.5	69.5
X_I	$g\ COD \cdot m^{-3}$	102.4	150
X_S	$g\ COD \cdot m^{-3}$	404.64	600
$X_{B,H}$	$g\ COD \cdot m^{-3}$	56.34	100
$X_{B,A}$	$g\ COD \cdot m^{-3}$	0	0
X_P	$g\ COD \cdot m^{-3}$	0	0
S_O	$g\ O_2 \cdot m^{-3}$	0	0
S_{NO}	$g\ N \cdot m^{-3}$	0	0
S_{NH}	$g\ N \cdot m^{-3}$	31.56	31.56
S_{ND}	$g\ N \cdot m^{-3}$	6.95	6.95

(*Continued*)

Table B.14 Influent composition of the two ring-test cases for the primary clarifier (*Continued*).

Variable	Unit	Case 1	Case 2
X_{ND}	$g\ N \cdot m^{-3}$	21.18	40
S_{ALK}	$mol\ HCO_3 \cdot m^{-3}$	7	7
TSS	$g\ SS \cdot m^{-3}$	422.535	637.5
Q_i	$m^3 \cdot d^{-1}$	18 446	18 446
T	°C	15	15

Table B.15 Effluent concentrations from primary clarifier for case 1.

Variable	Unit	GPS-X	FORTRAN	MATLAB	SIMBA	WEST	Average
S_I	$g\ COD \cdot m^{-3}$	30	30	30	30	30	30
S_S	$g\ COD \cdot m^{-3}$	69.5	69.5	69.5	69.5	69.5	69.5
X_I	$g\ COD \cdot m^{-3}$	52.04	52.01	52.04	52.04	52.04	52.04
X_S	$g\ COD \cdot m^{-3}$	205.6	205.6	205.6	205.6	205.6	205.6
$X_{B,H}$	$g\ COD \cdot m^{-3}$	28.63	28.63	28.63	28.63	28.63	28.63
$X_{B,A}$	$g\ COD \cdot m^{-3}$	0	0	0	0	0	0
X_P	$g\ COD \cdot m^{-3}$	0	0	0	0	0	0
S_O	$g\ O_2 \cdot m^{-3}$	0	0	0	0	0	0
S_{NO}	$g\ N \cdot m^{-3}$	0	0	0	0	0	0
S_{NH}	$g\ N \cdot m^{-3}$	31.56	31.56	31.56	31.56	31.56	31.56
S_{ND}	$g\ N \cdot m^{-3}$	6.95	6.95	6.95	6.95	6.95	6.95
X_{ND}	$g\ N \cdot m^{-3}$	10.76	10.76	10.76	10.76	10.76	10.76
S_{ALK}	$mol\ HCO_3 \cdot m^{-3}$	7.00	7.00	7.00	7.00	7.00	7.00
TSS	$g\ SS \cdot m^{-3}$	214.7	214.7	214.7	214.7	214.7	214.7
Q_e	$m^3 \cdot d^{-1}$	18 320	18 320	18 320	18 320	18 320	18 320

Table B.16 Underflow sludge concentrations from primary clarifier for case 1.

Variable	Unit	GPS-X	FORTRAN	MATLAB	SIMBA	WEST	Average
S_I	$g\ COD \cdot m^{-3}$	30	30	30	30	30	30
S_S	$g\ COD \cdot m^{-3}$	69.5	69.5	69.5	69.5	69.5	69.5
X_I	$g\ COD \cdot m^{-3}$	7 246	7 246	7 246	7 246	7 246	7 246
X_S	$g\ COD \cdot m^{-3}$	28 630	28 630	28 630	28 630	28 630	28 630
$X_{B,H}$	$g\ COD \cdot m^{-3}$	3 987	3 987	3 987	3 987	3 987	3 987
$X_{B,A}$	$g\ COD \cdot m^{-3}$	0	0	0	0	0	0
X_P	$g\ COD \cdot m^{-3}$	0	0	0	0	0	0
S_O	$g\ O_2 \cdot m^{-3}$	0	0	0	0	0	0

(*Continued*)

Table B.16 Underflow sludge concentrations from primary clarifier for case 1 (*Continued*).

Variable	Unit	GPS-X	FORTRAN	MATLAB	SIMBA	WEST	Average
S_{NO}	$g\ N \cdot m^{-3}$	0	0	0	0	0	0
S_{NH}	$g\ N \cdot m^{-3}$	31.56	31.56	31.56	31.56	31.56	31.56
S_{ND}	$g\ N \cdot m^{-3}$	6.95	6.95	6.95	6.95	6.95	6.95
X_{ND}	$g\ N \cdot m^{-3}$	1 499	1 499	1 499	1 499	1 499	1 499
S_{ALK}	$mol\ HCO_3 \cdot m^{-3}$	7	7	7	7	7	7
TSS	$g\ SS \cdot m^{-3}$	29 900	29 900	29 900	29 900	29 900	29 900
Q_e	$m^3 \cdot d^{-1}$	129.1	129.1	129.1	129.1	129.1	129.1

Table B.17 Effluent concentrations from primary clarifier for case 2.

Variable	Unit	GPS-X	FORTRAN	MATLAB	SIMBA	WEST	Average
S_I	$g\ COD \cdot m^{-3}$	30	30	30	30	30	30
S_S	$g\ COD \cdot m^{-3}$	69.5	69.5	69.5	69.5	69.5	69.5
X_I	$g\ COD \cdot m^{-3}$	76.23	76.23	76.23	76.23	76.23	76.23
X_S	$g\ COD \cdot m^{-3}$	304.9	304.9	304.9	304.9	304.9	304.9
$X_{B,H}$	$g\ COD \cdot m^{-3}$	50.82	50.82	50.82	50.82	50.82	50.82
$X_{B,A}$	$g\ COD \cdot m^{-3}$	0	0	0	0	0	0
X_P	$g\ COD \cdot m^{-3}$	0	0	0	0	0	0
S_O	$g\ O_2 \cdot m^{-3}$	0	0	0	0	0	0
S_{NO}	$g\ N \cdot m^{-3}$	0	0	0	0	0	0
S_{NH}	$g\ N \cdot m^{-3}$	31.56	31.56	31.56	31.56	31.56	31.56
S_{ND}	$g\ N \cdot m^{-3}$	6.95	6.95	6.95	6.95	6.95	6.95
X_{ND}	$g\ N \cdot m^{-3}$	20.33	20.33	20.33	20.33	20.33	20.33
S_{ALK}	$mol\ HCO_3 \cdot m^{-3}$	7	7	7	7	7	7
TSS	$g\ SS \cdot m^{-3}$	324	324	324	324	324	324
Q_e	$m^3 \cdot d^{-1}$	18 320	18 320	18 320	18 320	18 320	18 320

Table B.18 Underflow sludge concentrations from primary clarifier for case 2.

Variable	Unit	GPS-X	FORTRAN	MATLAB	SIMBA	WEST	Average
S_I	$g\ COD \cdot m^{-3}$	30	30	30	30	30	30
S_S	$g\ COD \cdot m^{-3}$	69.5	69.5	69.5	69.5	69.5	69.5
X_I	$g\ COD \cdot m^{-3}$	10 610	10 610	10 610	10 610	10 610	10 610
X_S	$g\ COD \cdot m^{-3}$	42 460	42 460	42 460	42 460	42 460	42 460
$X_{B,H}$	$g\ COD \cdot m^{-3}$	7 076	7 076	7 076	7 076	7 076	7 076
$X_{B,A}$	$g\ COD \cdot m^{-3}$	0	0	0	0	0	0
X_P	$g\ COD \cdot m^{-3}$	0	0	0	0	0	0
S_O	$g\ O_2 \cdot m^{-3}$	0	0	0	0	0	0

(*Continued*)

Table B.18 Underflow sludge concentrations from primary clarifier for case 2 (*Continued*).

Variable	Unit	GPS-X	FORTRAN	MATLAB	SIMBA	WEST	Average
S_{NO}	$g\ N \cdot m^{-3}$	0	0	0	0	0	0
S_{NH}	$g\ N \cdot m^{-3}$	31.56	31.56	31.56	31.56	31.56	31.56
S_{ND}	$g\ N \cdot m^{-3}$	6.95	6.95	6.95	6.95	6.95	6.95
X_{ND}	$g\ N \cdot m^{-3}$	2 831	2 831	2 831	2 831	2 831	2 831
S_{ALK}	$mol\ HCO_3 \cdot m^{-3}$	7	7	7	7	7	7
TSS	$g\ SS \cdot m^{-3}$	45 110	45 110	45 110	45 110	45 110	45 130
Q_e	$m^3 \cdot d^{-1}$	129.1	129.1	129.1	129.1	129.1	129.1

Sludge thickener

The sludge thickener was tested for the two influent cases shown in Table B.19. Identical results are found for the different implementations.

Table B.19 Influent composition of the two ring-test cases for the sludge thickener.

Variable	Unit	Case 1	Case 2
S_I	$g\ COD \cdot m^{-3}$	31.11	50
S_S	$g\ COD \cdot m^{-3}$	0.78	0.3
X_I	$g\ COD \cdot m^{-3}$	3 024.2	2 000
X_S	$g\ COD \cdot m^{-3}$	112.85	500
$X_{B,H}$	$g\ COD \cdot m^{-3}$	6 704	9 000
$X_{B,A}$	$g\ COD \cdot m^{-3}$	537.14	1 000
X_P	$g\ COD \cdot m^{-3}$	1 507.89	1 000
S_O	$g\ O_2 \cdot m^{-3}$	2.13	0.5
S_{NO}	$g\ N \cdot m^{-3}$	20.76	25
S_{NH}	$g\ N \cdot m^{-3}$	0.74	1
S_{ND}	$g\ N \cdot m^{-3}$	0.64	2
X_{ND}	$g\ N \cdot m^{-3}$	8.29	50
S_{ALK}	$mol\ HCO_3 \cdot m^{-3}$	25.92	10
TSS	$g\ SS \cdot m^{-3}$	8 914.56	10 125
Q_i	$m^3 \cdot d^{-1}$	300	500
T	°C	15	15

Table B.20 Effluent concentrations from sludge thickener for case 1.

Variable	Unit	GPS-X	FORTRAN	MATLAB	SIMBA	WEST	Average
S_I	$g\ COD \cdot m^{-3}$	31.11	31.11	31.11	31.11	31.11	31.11
S_S	$g\ COD \cdot m^{-3}$	0.78	0.78	0.78	0.78	0.78	0.78
X_I	$g\ COD \cdot m^{-3}$	69.11	69.11	69.11	69.11	69.11	69.11

(*Continued*)

Table B.20 Effluent concentrations from sludge thickener for case 1 (*Continued*).

Variable	Unit	GPS-X	FORTRAN	MATLAB	SIMBA	WEST	Average
X_S	$g\ COD \cdot m^{-3}$	2.579	2.579	2.579	2.579	2.579	2.579
$X_{B,H}$	$g\ COD \cdot m^{-3}$	153.2	153.2	153.2	153.2	153.2	153.2
$X_{B,A}$	$g\ COD \cdot m^{-3}$	12.27	12.27	12.27	12.27	12.27	12.27
X_P	$g\ COD \cdot m^{-3}$	34.46	34.46	34.46	34.46	34.46	34.46
S_O	$g\ O_2 \cdot m^{-3}$	2.13	2.13	2.13	2.13	2.13	2.13
S_{NO}	$g\ N \cdot m^{-3}$	20.76	20.76	20.76	20.76	20.76	20.76
S_{NH}	$g\ N \cdot m^{-3}$	0.74	0.74	0.74	0.74	0.74	0.74
S_{ND}	$g\ N \cdot m^{-3}$	0.64	0.64	0.64	0.64	0.64	0.64
X_{ND}	$g\ N \cdot m^{-3}$	0.1894	0.1894	0.1894	0.1894	0.1894	0.1894
S_{ALK}	$mol\ HCO_3 \cdot m^{-3}$	25.92	25.92	25.92	25.92	25.92	25.92
TSS	$g\ SS \cdot m^{-3}$	203.7	203.7	203.7	203.7	203.7	203.7
Q_e	$m^3 \cdot d^{-1}$	262.6	262.6	262.6	262.6	262.6	262.6

Table B.21 Underflow sludge concentrations from sludge thickener for case 1.

Variable	Unit	GPS-X	FORTRAN	MATLAB	SIMBA	WEST	Average
S_I	$g\ COD \cdot m^{-3}$	31.11	31.11	31.11	31.11	31.11	31.11
S_S	$g\ COD \cdot m^{-3}$	0.78	0.78	0.78	0.78	0.78	0.78
X_I	$g\ COD \cdot m^{-3}$	23 750	23 750	23 750	23 750	23 750	23 750
X_S	$g\ COD \cdot m^{-3}$	886.1	886.1	886.1	886.1	886.1	886.1
$X_{B,H}$	$g\ COD \cdot m^{-3}$	52 640	52 640	52 640	52 640	52 640	52 640
$X_{B,A}$	$g\ COD \cdot m^{-3}$	4 218	4 218	4 218	4 218	4 218	4 218
X_P	$g\ COD \cdot m^{-3}$	11 840	11 840	11 840	11 840	11 840	11 840
S_O	$g\ O_2 \cdot m^{-3}$	2.13	2.13	2.13	2.13	2.13	2.13
S_{NO}	$g\ N \cdot m^{-3}$	20.76	20.76	20.76	20.76	20.76	20.76
S_{NH}	$g\ N \cdot m^{-3}$	0.74	0.74	0.74	0.74	0.74	0.74
S_{ND}	$g\ N \cdot m^{-3}$	0.64	0.64	0.64	0.64	0.64	0.64
X_{ND}	$g\ N \cdot m^{-3}$	65.1	65.1	65.1	65.1	65.1	65.1
S_{ALK}	$mol\ HCO_3 \cdot m^{-3}$	25.92	25.92	25.92	25.92	25.92	25.92
TSS	$g\ SS \cdot m^{-3}$	70 000	70 000	70 000	70 000	70 000	70 000
Q_u	$m^3 \cdot d^{-1}$	37.44	37.44	37.44	37.44	37.44	37.44

Table B.22 Effluent concentrations from sludge thickener for case 2.

Variable	Unit	GPS-X	FORTRAN	MATLAB	SIMBA	WEST	Average
S_I	$g\ COD \cdot m^{-3}$	50	50	50	50	50	50
S_S	$g\ COD \cdot m^{-3}$	0.3	0.3	0.3	0.3	0.3	0.3
X_I	$g\ COD \cdot m^{-3}$	46.61	46.61	46.61	46.61	46.61	46.61

(*Continued*)

Table B.22 Effluent concentrations from sludge thickener for case 2 (*Continued*).

Variable	Unit	GPS-X	FORTRAN	MATLAB	SIMBA	WEST	Average
X_S	g COD·m^{-3}	11.65	11.65	11.65	11.65	11.65	11.65
$X_{B,H}$	g COD·m^{-3}	209.7	209.7	209.7	209.7	209.7	209.7
$X_{B,A}$	g COD·m^{-3}	23.3	23.3	23.3	23.3	23.3	23.3
X_P	g COD·m^{-3}	23.3	23.3	23.3	23.3	23.3	23.3
S_O	g O_2·m^{-3}	0.5	0.5	0.5	0.5	0.5	0.5
S_{NO}	g N·m^{-3}	25	25	25	25	25	25
S_{NH}	g N·m^{-3}	1	1	1	1	1	1
S_{ND}	g N·m^{-3}	2	2	2	2	2	2
X_{ND}	g N·m^{-3}	1.165	1.165	1.165	1.165	1.165	1.165
S_{ALK}	mol HCO_3·m^{-3}	10	10	10	10	10	10
TSS	g SS·m^{-3}	235.9	235.9	235.9	235.9	235.9	235.9
Q_e	m^3·d^{-1}	429.1	429.1	429.1	429.1	429.1	429.1

Table B.23 Underflow sludge concentrations from sludge thickener for case 2.

Variable	Unit	GPS-X	FORTRAN	MATLAB	SIMBA	WEST	Average
S_I	g COD·m^{-3}	50	50	50	50	50	50
S_S	g COD·m^{-3}	0.3	0.3	0.3	0.3	0.3	0.3
X_I	g COD·m^{-3}	13 830	13 830	13 830	13 830	13 830	13 830
X_S	g COD·m^{-3}	3 457	3 457	3 457	3 457	3 457	3 457
$X_{B,H}$	g COD·m^{-3}	62 220	62 220	62 220	62 220	62 220	62 220
$X_{B,A}$	g COD·m^{-3}	6 914	6 914	6 914	6 914	6 914	6 914
X_P	g COD·m^{-3}	6 914	6 914	6 914	6 914	6 914	6 914
S_O	g O_2·m^{-3}	0.5	0.5	0.5	0.5	0.5	0.5
S_{NO}	g N·m^{-3}	25	25	25	25	25	25
S_{NH}	g N·m^{-3}	1	1	1	1	1	1
S_{ND}	g N·m^{-3}	2	2	2	2	2	2
X_{ND}	g N·m^{-3}	345.7	345.7	345.7	345.7	345.7	345.7
S_{ALK}	mol HCO_3·m^{-3}	10	10	10	10	10	10
TSS	g SS·m^{-3}	70 000	70 000	70 000	70 000	70 000	70 000
Q_u	m^3·d^{-1}	70.88	70.87	70.88	70.88	70.88	70.88

ASM1 to ADM1 interface

The ASM1 to ADM1 interface was tested for two influent cases, one reflecting primary sludge and one reflecting secondary sludge. Compositions are shown in Table B.24.

Table B.24 Influent composition of the two ring-test cases for the ASM1 to ADM1 interface.

Variable	Unit	Case 1	Case 2
S_I	$g\ COD \cdot m^{-3}$	31.11	50
S_S	$g\ COD \cdot m^{-3}$	0.78	70
X_I	$g\ COD \cdot m^{-3}$	23 800	7 200
X_S	$g\ COD \cdot m^{-3}$	890	28 700
$X_{B,H}$	$g\ COD \cdot m^{-3}$	52 000	4 000
$X_{B,A}$	$g\ COD \cdot m^{-3}$	4 200	0
X_P	$g\ COD \cdot m^{-3}$	11 900	0
S_O	$g\ O_2 \cdot m^{-3}$	2	0
S_{NO}	$g\ N \cdot m^{-3}$	20	0
S_{NH}	$g\ N \cdot m^{-3}$	0.74	31
S_{ND}	$g\ N \cdot m^{-3}$	0.64	7
X_{ND}	$g\ N \cdot m^{-3}$	65	1 500
S_{ALK}	$mol\ HCO_3 \cdot m^{-3}$	8	7
TSS	$g\ SS \cdot m^{-3}$	69 592.5	29 925
Q_i	$m^3 \cdot d^{-1}$	37	130
T	°C	15	15

Results were found to be identical for all implementations.

Table B.25 Output concentrations from ASM1 to ADM1 interface for case 1 (only non-zero states are shown).

Variable	Unit	FORTRAN	MATLAB	SIMBA	WEST	Average
S_{su}	$kg\ COD \cdot m^{-3}$	0.008	0.008	0.008	0.008	0.008
S_{IC}	$kmol\ C \cdot m^{-3}$	0.01127	0.01127	0.01127	0.01127	0.01127
S_I	$kg\ COD \cdot m^{-3}$	0.023	0.023	0.023	0.023	0.023
X_{ch}	$kg\ COD \cdot m^{-3}$	2.06	2.06	2.06	2.06	2.06
X_{pr}	$kg\ COD \cdot m^{-3}$	35.53	35.53	35.53	35.53	35.53
X_{li}	$kg\ COD \cdot m^{-3}$	1.457	1.457	1.457	1.457	1.457
X_I	$kg\ COD \cdot m^{-3}$	53.68	53.68	53.68	53.68	53.68
S_{an}	$kmol \cdot m^{-3}$	0.009375	0.009376	0.009376	0.009376	0.09376

Table B.26 Output concentrations from ASM1 to ADM1 interface for case 2 (only non-zero states are shown).

Variable	Unit	FORTRAN	MATLAB	SIMBA	WEST	Average
S_{aa}	$kg\ COD \cdot m^{-3}$	0.07	0.07	0.07	0.07	0.07
S_{IC}	$kmol\ C \cdot m^{-3}$	0.008248	0.008248	0.008248	0.008248	0.008248

(*Continued*)

Table B.26 Output concentrations from ASM1 to ADM1 interface for case 2 (only non-zero states are shown) (*Continued*).

Variable	Unit	FORTRAN	MATLAB	SIMBA	WEST	Average
S_{IN}	kmol N $\cdot$ m^{-3}	0.002096	0.002096	0.002096	0.002096	0.002096
S_I	kg COD $\cdot$ m^{-3}	0.03	0.03	0.03	0.03	0.03
X_{ch}	kg COD $\cdot$ m^{-3}	4.161	4.161	4.161	4.161	4.161
X_{pr}	kg COD $\cdot$ m^{-3}	17.79	17.79	17.79	17.79	17.79
X_{li}	kg COD $\cdot$ m^{-3}	9.471	9.471	9.471	9.471	9.471
X_I	kg COD $\cdot$ m^{-3}	8.48	8.48	8.48	8.48	8.48
S_{an}	kmol $\cdot$ m^{-3}	0.004786	0.004786	0.004786	0.004786	0.004786

Anaerobic digester

The anaerobic digester was tested using two different influent cases. The compositions used are shown in Table B.27.

Table B.27 Influent composition of the two ring-test cases for the anaerobic digester.

Variable	Unit	Case 1	Case 2
S_{su}	kg COD $\cdot$ m^{-3}	0	0.01
S_{aa}	kg COD $\cdot$ m^{-3}	0.05	0.001
S_{fa}	kg COD $\cdot$ m^{-3}	0	0.001
S_{va}	kg COD $\cdot$ m^{-3}	0	0.001
S_{bu}	kg COD $\cdot$ m^{-3}	0	0.001
S_{pro}	kg COD $\cdot$ m^{-3}	0	0.001
S_{ac}	kg COD $\cdot$ m^{-3}	0	0.001
S_{h2}	kg COD $\cdot$ m^{-3}	0	$1.00 \cdot 10^{-8}$
S_{ch4}	kg COD $\cdot$ m^{-3}	0	0.001
S_{IC}	kmol C $\cdot$ m^{-3}	0.006	0.005
S_{IN}	kmol N $\cdot$ m^{-3}	0.07	0.1
S_I	kg COD $\cdot$ m^{-3}	0.06	0.1
X_c	kg COD $\cdot$ m^{-3}	37	5
X_{ch}	kg COD $\cdot$ m^{-3}	0	4
X_{pr}	kg COD $\cdot$ m^{-3}	0	30
X_{li}	kg COD $\cdot$ m^{-3}	0	3
X_{su}	kg COD $\cdot$ m^{-3}	0	0.1
X_{aa}	kg COD $\cdot$ m^{-3}	0	0.1
X_{fa}	kg COD $\cdot$ m^{-3}	0	0.1
X_{c4}	kg COD $\cdot$ m^{-3}	0	0.1
X_{pro}	kg COD $\cdot$ m^{-3}	0	0.1
X_{ac}	kg COD $\cdot$ m^{-3}	0	0.1

(*Continued*)

Table B.27 Influent composition of the two ring test cases for the anaerobic digester (*Continued*).

Variable	Unit	Case 1	Case 2
X_{h2}	kg COD $\cdot$ m^{-3}	0	0.1
X_I	kg COD $\cdot$ m^{-3}	12	30
S_{cat}	kmol $\cdot$ m^{-3}	0.006	0.005
S_{an}	kmol $\cdot$ m^{-3}	0.07	0.26
Q_i	m$^3 \cdot$ d^{-1}	166	170
T	°C	35	40

Results are summarised in the following tables.

Table B.28 Output concentrations and biogas related variables from anaerobic digester for case 1.

Variable	Unit	FORTRAN	MATLAB	SIMBA	WEST	Average
S_{su}	kg COD $\cdot$ m^{-3}	0.01176	0.01175	0.01175	0.01175	0.01175
S_{aa}	kg COD $\cdot$ m^{-3}	0.005259	0.005258	0.005258	0.005258	0.005258
S_{fa}	kg COD $\cdot$ m^{-3}	0.1003	0.1003	0.1003	0.1003	0.1003
S_{va}	kg COD $\cdot$ m^{-3}	0.01058	0.01056	0.01056	0.01056	0.01056
S_{bu}	kg COD $\cdot$ m^{-3}	0.014	0.01397	0.01397	0.01397	0.01398
S_{pro}	kg COD $\cdot$ m^{-3}	0.01648	0.01648	0.01648	0.01648	0.01648
S_{ac}	kg COD $\cdot$ m^{-3}	0.05486	0.05594	0.05596	0.05594	0.05568
S_{h2}	kg COD $\cdot$ m^{-3}	$2.376 \cdot 10^{-7}$	$2.374 \cdot 10^{-7}$	$2.374 \cdot 10^{-7}$	$2.374 \cdot 10^{-7}$	$2.347 \cdot 10^{-7}$
S_{ch4}	kg COD $\cdot$ m^{-3}	0.05289	0.05242	0.05233	0.05242	0.05252
S_{IC}	kmol C $\cdot$ m^{-3}	0.04575	0.0461	0.0461	0.0461	0.04601
S_{IN}	kmol N $\cdot$ m^{-3}	0.1021	0.1022	0.1022	0.1022	0.1022
S_I	kg COD $\cdot$ m^{-3}	3.52	3.52	3.52	3.52	3.52
X_c	kg COD $\cdot$ m^{-3}	3.379	3.379	3.379	3.379	3.379
X_{ch}	kg COD $\cdot$ m^{-3}	0.03362	0.03362	0.03362	0.03362	0.03362
X_{pr}	kg COD $\cdot$ m^{-3}	0.03362	0.03362	0.03362	0.03362	0.03362
X_{li}	kg COD $\cdot$ m^{-3}	0.05044	0.05044	0.05044	0.05044	0.05044
X_{su}	kg COD $\cdot$ m^{-3}	0.5242	0.5244	0.5244	0.5244	0.5243
X_{aa}	kg COD $\cdot$ m^{-3}	0.3932	0.3934	0.3934	0.3934	0.3933
X_{fa}	kg COD $\cdot$ m^{-3}	0.4133	0.4134	0.4134	0.4134	0.4134
X_{c4}	kg COD $\cdot$ m^{-3}	0.1685	0.1688	0.1688	0.1688	0.1687
X_{pro}	kg COD $\cdot$ m^{-3}	0.0805	0.08052	0.08052	0.08052	0.08052
X_{ac}	kg COD $\cdot$ m^{-3}	0.5487	0.549	0.549	0.549	0.549
X_{h2}	kg COD $\cdot$ m^{-3}	0.2652	0.2653	0.2653	0.2653	0.2653
X_I	kg COD $\cdot$ m^{-3}	18.92	18.92	18.92	18.92	18.92
S_{cat}	kmol $\cdot$ m^{-3}	0.006	0.006	0.006	0.006	0.006
S_{an}	kmol $\cdot$ m^{-3}	0.07	0.07	0.07	0.07	0.07

(*Continued*)

Table B.28 Output concentrations and biogas related variables from anaerobic digester for case 1 (*Continued*).

Variable	Unit	FORTRAN	MATLAB	SIMBA	WEST	Average
Q_{out} (water)	$m^3 \cdot d^{-1}$	166	166	166	166	166
T	°C	35	35	35	35	35
pH	–	6.88	6.87	6.87	6.87	6.872
$S_{gas,h2}$	kg COD $\cdot m^{-3}$	0.00001078	0.00001076	n/a	0.00001076	0.00001077
$S_{gas,ch4}$	kg COD $\cdot m^{-3}$	1.609	1.593	n/a	1.593	1.598
$S_{gas,co2}$	kg COD $\cdot m^{-3}$	0.0138	0.01415	n/a	0.01415	0.01403
$p_{gas,h2}$	bar	0.00001727	0.00001723	0.0000172	0.00001723	0.00001723
$p_{gas,ch4}$	bar	0.6439	0.6375	0.6375	0.6375	0.6391
$p_{gas,co2}$	bar	0.3536	0.3625	0.3625	0.3625	0.3603
$p_{gas,tot}$	bar	1.054	1.056	1.056	1.056	1.056
Q_{out} (gas)	N $m^3 \cdot d^{-1}$	2 201.0	2 228.3	2 227.9	2 142.0	2 199.8

Table B.29 Output concentrations and biogas related variables from anaerobic digester for case 2.

Variable	Unit	FORTRAN	MATLAB	SIMBA	WEST	Average
S_{su}	kg COD $\cdot m^{-3}$	0.01008	0.01007	0.01007	0.01007	0.01007
S_{aa}	kg COD $\cdot m^{-3}$	0.005136	0.005136	0.005136	0.005136	0.005136
S_{fa}	kg COD $\cdot m^{-3}$	0.06976	0.06969	0.06969	0.06969	0.06971
S_{va}	kg COD $\cdot m^{-3}$	0.01068	0.01067	0.01067	0.01067	0.01067
S_{bu}	kg COD $\cdot m^{-3}$	0.01186	0.01186	0.01186	0.01186	0.01186
S_{pro}	kg COD $\cdot m^{-3}$	0.01124	0.01124	0.01124	0.01124	0.01124
S_{ac}	kg COD $\cdot m^{-3}$	0.1267	0.1246	0.1293	0.1246	0.1263
S_{h2}	kg COD $\cdot m^{-3}$	$2.024 \cdot 10^{-7}$	$2.024 \cdot 10^{-7}$	$2.024 \cdot 10^{-7}$	$2.024 \cdot 10^{-7}$	$2.024 \cdot 10^{-7}$
S_{ch4}	kg COD $\cdot m^{-3}$	0.04726	0.0474	0.05100	0.0474	0.04826
S_{IC}	kmol C $\cdot m^{-3}$	0.03826	0.03933	0.04085	0.03933	0.03944
S_{IN}	kmol N $\cdot m^{-3}$	0.287	0.2876	0.2876	0.2876	0.2875
S_I	kg COD $\cdot m^{-3}$	0.7368	0.7378	0.7378	0.7378	0.7376
X_c	kg COD $\cdot m^{-3}$	0.6377	0.6378	0.6378	0.6378	0.6378
X_{ch}	kg COD $\cdot m^{-3}$	0.02625	0.02625	0.02625	0.02625	0.02625
X_{pr}	kg COD $\cdot m^{-3}$	0.1556	0.1556	0.1556	0.1556	0.1556
X_{li}	kg COD $\cdot m^{-3}$	0.02444	0.02444	0.02444	0.02444	0.02444
X_{su}	kg COD $\cdot m^{-3}$	0.4637	0.4638	0.4638	0.4638	0.4638
X_{aa}	kg COD $\cdot m^{-3}$	1.849	1.849	1.849	1.849	1.849
X_{fa}	kg COD $\cdot m^{-3}$	0.2673	0.2675	0.2675	0.2675	0.2675
X_{c4}	kg COD $\cdot m^{-3}$	0.699	0.6993	0.6993	0.6993	0.6992
X_{pro}	kg COD $\cdot m^{-3}$	0.2453	0.2455	0.2455	0.2455	0.2455
X_{ac}	kg COD $\cdot m^{-3}$	1.059	1.06	1.06	1.06	1.06

(Continued)

Table B.29 Output concentrations and biogas related variables from anaerobic digester for case 2 (*Continued*).

Variable	Unit	FORTRAN	MATLAB	SIMBA	WEST	Average
X_{h2}	kg COD · m^{-3}	0.4531	0.4532	0.4532	0.4532	0.4531
X_I	kg COD · m^{-3}	31.27	31.28	31.28	31.28	31.28
S_{cat}	kmol · m^{-3}	0.005	0.005	0.005	0.005	0.005
S_{an}	kmol · m^{-3}	0.26	0.26	0.26	0.26	0.26
Q_{out} (water)	m^3 · d^{-1}	170	170	170	170	170
pH	–	6.68	6.694	6.646	6.694	6.68
$S_{gas,h2}$	kg COD · m^{-3}	$8.115 \cdot 10^{-6}$	$8.121 \cdot 10^{-6}$	n.a.	$8.121 \cdot 10^{-6}$	$8.119 \cdot 10^{-6}$
$S_{gas,ch4}$	kg COD · m^{-3}	1.395	1.4	n.a.	1.4	1.398
$S_{gas,co2}$	kg COD · m^{-3}	0.01729	0.01746	n.a.	0.01746	0.0174
$p_{gas,h2}$	bar	0.00001321	0.00001322	0.00001297	0.00001322	0.00001316
$p_{gas,ch4}$	bar	0.5673	0.5694	0.5704	0.5694	0.5691
$p_{gas,co2}$	bar	0.4501	0.4545	0.4538	0.4545	0.4532
$p_{gas,tot}$	bar	1.091	1.097	1.097	1.097	1.096
Q_{out} (gas)	N m^3 · d^{-1}	4 534	4 556	4 545	4 556	4 548

ADM1 to ASM1 interface

The ADM1 to ASM1 interface was tested using two cases. The influent composition was equivalent in both cases (Table B.30) except that the digester's pH was 7.0 in case 1 and 8.0 in case 2.

Table B.30 Influent composition of the two ring-test cases for the ADM1 to ASM1 interface.

Variable	Unit	Case 1 and 2
S_{su}	kg COD · m^{-3}	0.01247
S_{aa}	kg COD · m^{-3}	0.00558
S_{fa}	kg COD · m^{-3}	0.10825
S_{va}	kg COD · m^{-3}	0.01123
S_{bu}	kg COD · m^{-3}	0.01489
S_{pro}	kg COD · m^{-3}	0.01771
S_{ac}	kg COD · m^{-3}	0.09187
S_{h2}	kg COD · m^{-3}	0
S_{ch4}	kg COD · m^{-3}	0.04982
S_{IC}	kmol C · m^{-3}	0.00977
S_{IN}	kmol N · m^{-3}	0.09325
S_I	kg COD · m^{-3}	3.20098
X_c	kg COD · m^{-3}	3.34033
X_{ch}	kg COD · m^{-3}	0.03323

(*Continued*)

Table B.30 Influent composition of the two ring-test cases for the ADM1 to ASM1 interface (*Continued*).

Variable	Unit	Case 1 and 2
X_{pr}	kg COD·m^{-3}	0.03323
X_{li}	kg COD·m^{-3}	0.04984
X_{su}	kg COD·m^{-3}	0.48952
X_{aa}	kg COD·m^{-3}	0.36483
X_{fa}	kg COD·m^{-3}	0.38536
X_{c4}	kg COD·m^{-3}	0.15657
X_{pro}	kg COD·m^{-3}	0.07488
X_{ac}	kg COD·m^{-3}	0.50955
X_{h2}	kg COD·m^{-3}	0.2471
X_I	kg COD·m^{-3}	18.44098
S_{cat}	kmol·m^{-3}	0
S_{an}	kmol·m^{-3}	0
Q_i	m^3·d^{-1}	100
T	°C	35

Results are identical for all implementations and summarised in Tables B.31 and B.32.

Table B.31 Output concentrations from ADM1 to ASM1 interface for case 1.

Variable	Unit	FORTRAN	MATLAB	SIMBA	WEST	Average
S_I	g COD·m^{-3}	3 201	3 201	3 201	3 201	3 201
S_S	g COD·m^{-3}	262	262	262	262	262
X_I	g COD·m^{-3}	18 440	18 440	18 440	18 440	18 440
X_S	g COD·m^{-3}	5 217	5 217	5 217	5 217	5 217
$X_{B,H}$	g COD·m^{-3}	0	0	0	0	0
$X_{B,A}$	g COD·m^{-3}	0	0	0	0	0
X_P	g COD·m^{-3}	467.8	467.8	467.8	467.8	467.8
S_O	g O_2·m^{-3}	0	0	0	0	0
S_{NO}	g N·m^{-3}	0	0	0	0	0
S_{NH}	g N·m^{-3}	1 582	1 582	1 582	1 582	1 582
S_{ND}	g N·m^{-3}	0.5468	0.5468	0.5468	0.5468	0.5468
X_{ND}	g N·m^{-3}	195	195	195	195	195
S_{ALK}	mol HCO_3·m^{-3}	30.59	30.6	30.6	30.6	30.6
TSS	g SS·m^{-3}	18 090	18 090	18 090	18 090	18 090

Table B.32 Output concentrations from ADM1 to ASM1 interface for case 2.

Variable	Unit	FORTRAN	MATLAB	SIMBA	WEST	Average
S_I	$g\ COD \cdot m^{-3}$	3 201	3 201	3 201	3 201	3 201
S_S	$g\ COD \cdot m^{-3}$	262	262	262	262	262
X_I	$g\ COD \cdot m^{-3}$	18 440	18 440	18 440	18 440	18 440
X_S	$g\ COD \cdot m^{-3}$	5 217	5 217	5 217	5 217	5 217
$X_{B,H}$	$g\ COD \cdot m^{-3}$	0	0	0	0	0
$X_{B,A}$	$g\ COD \cdot m^{-3}$	0	0	0	0	0
X_P	$g\ COD \cdot m^{-3}$	467.8	467.8	467.8	467.8	467.8
S_O	$g\ O_2 \cdot m^{-3}$	0	0	0	0	0
S_{NO}	$g\ N \cdot m^{-3}$	0	0	0	0	0
S_{NH}	$g\ N \cdot m^{-3}$	1 582	1 582	1 582	1 582	1 582
S_{ND}	$g\ N \cdot m^{-3}$	0.5468	0.5468	0.5468	0.5468	0.5468
X_{ND}	$g\ N \cdot m^{-3}$	195	195	195	195	195
S_{ALK}	$mol\ HCO_3 \cdot m^{-3}$	40.35	40.35	40.35	40.35	40.35
TSS	$g\ SS \cdot m^{-3}$	18 090	18 090	18 090	18 090	18 090

Dewatering unit

The dewatering unit was tested using two influent cases, which are shown in Table B.33.

Table B.33 Influent composition of the two ring-test cases for the dewatering unit.

Variable	Unit	Case 1	Case 2
S_I	$g\ COD \cdot m^{-3}$	500	1 000
S_S	$g\ COD \cdot m^{-3}$	250	100
X_I	$g\ COD \cdot m^{-3}$	20 000	10 000
X_S	$g\ COD \cdot m^{-3}$	6 000	10 000
$X_{B,H}$	$g\ COD \cdot m^{-3}$	0	1 000
$X_{B,A}$	$g\ COD \cdot m^{-3}$	0	1 000
X_P	$g\ COD \cdot m^{-3}$	0	1 000
S_O	$g\ O_2 \cdot m^{-3}$	0	10
S_{NO}	$g\ N \cdot m^{-3}$	0	1 000
S_{NH}	$g\ N \cdot m^{-3}$	1 400	1 000
S_{ND}	$g\ N \cdot m^{-3}$	200	200
X_{ND}	$g\ N \cdot m^{-3}$	300	300
S_{ALK}	$mol\ HCO_3 \cdot m^{-3}$	7	50
TSS	$g\ SS \cdot m^{-3}$	19 500	17 250
Q_i	$m^3 \cdot d^{-1}$	165	100
T	°C	15	25

Results are summarised in the following tables.

Table B.34 Effluent concentrations from dewatering unit for case 1.

Variable	Unit	GPS-X	FORTRAN	MATLAB	SIMBA	WEST	Average
S_I	$g\ COD \cdot m^{-3}$	500	500	500	500	500	500
S_S	$g\ COD \cdot m^{-3}$	250	250	250	250	250	250
X_I	$g\ COD \cdot m^{-3}$	429.3	429.3	429.3	429.3	429.3	429.3
X_S	$g\ COD \cdot m^{-3}$	128.8	128.8	128.8	128.8	128.8	128.8
$X_{B,H}$	$g\ COD \cdot m^{-3}$	0	0	0	0	0	0
$X_{B,A}$	$g\ COD \cdot m^{-3}$	0	0	0	0	0	0
X_P	$g\ COD \cdot m^{-3}$	0	0	0	0	0	0
S_O	$g\ O_2 \cdot m^{-3}$	0	0	0	0	0	0
S_{NO}	$g\ N \cdot m^{-3}$	0	0	0	0	0	0
S_{NH}	$g\ N \cdot m^{-3}$	1 400	1 400	1 400	1 400	1 400	1 400
S_{ND}	$g\ N \cdot m^{-3}$	200	200	200	200	200	200
X_{ND}	$g\ N \cdot m^{-3}$	6.44	6.44	6.439	6.439	6.439	6.439
S_{ALK}	$mol\ HCO_3 \cdot m^{-3}$	7	7	7	7	7	7
TSS	$g\ SS \cdot m^{-3}$	418.6	418.6	418.6	418.6	418.6	418.6
Q_e	$m^3 \cdot d^{-1}$	153.7	153.7	153.7	153.7	153.7	153.7

Table B.35 Underflow sludge concentrations from dewatering unit for case 1.

Variable	Unit	GPS-X	FORTRAN	MATLAB	SIMBA	WEST	Average
S_I	$g\ COD \cdot m^{-3}$	500	500	500	500	500	500
S_S	$g\ COD \cdot m^{-3}$	250	250	250	250	250	250
X_I	$g\ COD \cdot m^{-3}$	287 200	287 200	287 200	287 200	287 200	287 200
X_S	$g\ COD \cdot m^{-3}$	86 150	86 150	86 150	86 150	86 150	86 150
$X_{B,H}$	$g\ COD \cdot m^{-3}$	0	0	0	0	0	0
$X_{B,A}$	$g\ COD \cdot m^{-3}$	0	0	0	0	0	0
X_P	$g\ COD \cdot m^{-3}$	0	0	0	0	0	0
S_O	$g\ O_2 \cdot m^{-3}$	0	0	0	0	0	0
S_{NO}	$g\ N \cdot m^{-3}$	0	0	0	0	0	0
S_{NH}	$g\ N \cdot m^{-3}$	1 400	1 400	1 400	1 400	1 400	1 400
S_{ND}	$g\ N \cdot m^{-3}$	200	200	200	200	200	200
X_{ND}	$g\ N \cdot m^{-3}$	4 308	4 308	4 308	4 308	4 308	4 308
S_{ALK}	$mol\ HCO_3 \cdot m^{-3}$	7	7	7	7	7	7
TSS	$g\ SS \cdot m^{-3}$	280 000	280 000	280 000	280 000	280 000	280 000
Q_u	$m^3 \cdot d^{-1}$	11.26	11.26	11.26	11.26	11.26	11.26

Table B.36 Effluent concentrations from dewatering unit for case 2.

Variable	Unit	GPS-X	FORTRAN	MATLAB	SIMBA	WEST	Average
S_I	$g\ COD \cdot m^{-3}$	1 000.00	1 000.00	1 000.00	1 000.00	1 000.00	1 000.00
S_S	$g\ COD \cdot m^{-3}$	100.00	100.00	100.00	100.00	100.00	100.00
X_I	$g\ COD \cdot m^{-3}$	212.90	212.90	212.90	212.90	212.90	212.90
X_S	$g\ COD \cdot m^{-3}$	212.90	212.90	212.90	212.90	212.90	212.90
$X_{B,H}$	$g\ COD \cdot m^{-3}$	21.29	21.29	21.29	21.29	21.29	21.29
$X_{B,A}$	$g\ COD \cdot m^{-3}$	21.29	21.29	21.29	21.29	21.29	21.29
X_P	$g\ COD \cdot m^{-3}$	21.29	21.29	21.29	21.29	21.29	21.29
S_O	$g\ O_2 \cdot m^{-3}$	10.00	10.00	10.00	10.00	10.00	10.00
S_{NO}	$g\ N \cdot m^{-3}$	1 000.00	1 000.00	1 000.00	1 000.00	1 000.00	1 000.00
S_{NH}	$g\ N \cdot m^{-3}$	1 000.00	1 000.00	1 000.00	1 000.00	1 000.00	1 000.00
S_{ND}	$g\ N \cdot m^{-3}$	200.00	200.00	200.00	200.00	200.00	200.00
X_{ND}	$g\ N \cdot m^{-3}$	6.39	6.39	6.39	6.39	6.39	6.39
S_{ALK}	$mol\ HCO_3 \cdot m^{-3}$	50.00	50.00	50.00	50.00	50.00	50.00
TSS	$g\ SS \cdot m^{-3}$	367.20	367.20	367.20	367.20	367.20	367.20
Q_e	$m^3 \cdot d^{-1}$	93.96	93.96	93.96	93.96	93.96	93.96

Table B.37 Underflow sludge concentrations from dewatering unit for case 2.

Variable	Unit	GPS-X	FORTRAN	MATLAB	SIMBA	WEST	Average
S_I	$g\ COD \cdot m^{-3}$	1 000	1 000	1 000	1 000	1 000	1 000
S_S	$g\ COD \cdot m^{-3}$	100	100	100	100	100	100
X_I	$g\ COD \cdot m^{-3}$	162 300	162 300	162 300	162 300	162 300	162 300
X_S	$g\ COD \cdot m^{-3}$	162 300	162 300	162 300	162 300	162 300	162 300
$X_{B,H}$	$g\ COD \cdot m^{-3}$	16 230	16 230	16 230	16 230	16 230	16 230
$X_{B,A}$	$g\ COD \cdot m^{-3}$	16 230	16 230	16 230	16 230	16 230	16 230
X_P	$g\ COD \cdot m^{-3}$	16 230	16 230	16 230	16 230	16 230	16 230
S_O	$g\ O_2 \cdot m^{-3}$	10	10	10	10	10	10
S_{NO}	$g\ N \cdot m^{-3}$	1 000	1 000	1 000	1 000	1 000	1 000
S_{NH}	$g\ N \cdot m^{-3}$	1 000	1 000	1 000	1 000	1 000	1 000
S_{ND}	$g\ N \cdot m^{-3}$	200	200	200	200	200	200
X_{ND}	$g\ N \cdot m^{-3}$	4 870	4 870	4 870	4 870	4 870	4 870
S_{ALK}	$mol\ HCO_3 \cdot m^{-3}$	50	50	50	50	50	50
TSS	$g\ SS \cdot m^{-3}$	280 000	280 000	280 000	280 000	280 000	280 000
Q_u	$m^3 \cdot d^{-1}$	6.04	6.04	6.04	6.04	6.04	6.04

Index

IWA Publishing's authorised EU representative for General Product Safety Regulations is Diane D'Arras, 15 rue Duret, 75116 Paris, France, e-mail: safety@iwap.co.uk.

Printed and bound by CPI Group (UK) Ltd, Croydon, CR0 4YY

13/05/2026

02109604-0013